Claims, Disputes and Litigation Involving BIM

Are you unsure about:

- the current US legal environment with respect to BIM and VDC?
- the evolving standards of care for design and construction professionals using BIM and VDC?
- what practical methods and techniques can be used for analyzing potential construction claims and disputes involving BIM technologies and VDC processes?

Building Information Modeling (BIM) technologies and Virtual Design and Construction (VDC) processes are aggressively and fundamentally changing the design, construction, and operation of buildings. Initial case studies of BIM have highlighted the potential these technologies have to reduce the need for claims, disputes and litigation, but evidence from other early sources shows they are not universally successful in this.

This timely and unique book provides crucial new methods for analyzing construction disputes in this emerging architecture, engineering, and construction (AEC) technological landscape. It explains how BIM and VDC can affect, and in some instances have already significantly altered the production and delivery of construction drawings, quantity surveys, and schedules, and how these changes might impact construction disputes.

The findings and advice in this book are indispensable to any stakeholder in a construction project using BIM. It will help contractors, cost managers, architects, building designers, quantity surveyors, and project managers to navigate and understand their responsibilities and exposure to risk when working with this new technology.

Jason M. Dougherty is a BIM/VDC specialist with over a decade of practical experience across the entire AEC industry spectrum. He has worked with some of the largest architectural firms in the world utilizing technology to drive solutions to architecture and engineering issues, assisted owners with successfully implementing BIM/VDC both contractually and technologically across their enterprises, and served as a retained expert in disputes involving BIM/VDC issues. He has had numerous articles on BIM/VDC published in trade and professional journals, lectured internationally, and has given courses on the same subject both to students and to practitioners.

Claims, Disputes and Litigation Involving BIM

Jason M. Dougherty

Routledge
Taylor & Francis Group

LONDON AND NEW YORK

First published 2015
by Routledge
2 Park Square, Milton Park, Abingdon, Oxon OX14 4RN

and by Routledge
711 Third Avenue, New York, NY 10017

Routledge is an imprint of the Taylor & Francis Group, an informa business

British Library Cataloguing-in-Publication Data
A catalogue record for this book is available from the British Library

Library of Congress Cataloging-in-Publication Data
Dougherty, Jason M., author.
Claims, disputes and litigation involving BIM / Jason M. Dougherty.
pages cm
Includes bibliographical references and index.
1. Construction contracts—United States. 2. Building information
modeling—United States. I. Title. II. Title: Claims, disputes and
litigation involving building information modeling.
KF902.D68 2015
343.7307′869—dc23
2015000646

ISBN: 978-0-415-85894-6 (hbk)
ISBN: 978-1-315-68901-2 (ebk)

Typeset in Goudy
by Swales & Willis Ltd, Exeter, Devon, UK

Contents

List of illustrations   viii
Preface   ix
Executive summary   x

PART I
Fundamentals: construction claims, and BIM/VDC   1

1   **Traditional construction claims**   3
    1.1. Contractor claims against owners 4
        1.1.1 Scope changes – overview 5
        1.1.2 Acceleration – overview 7
        1.1.3 Delay – overview 9
        1.1.4 Disruption – overview 11
    1.2. Owner claims against contractors 12
    1.3. Torts 12
    1.4. Summary 17

2   **General overview of BIM and VDC**   22
    2.1. General overview 22
    2.2. Preliminary definitions 25
    2.3. Design BIM – general outline of tools and workflows 26
    2.4. Design BIM – general outline of processes and procedures 27
    2.5. Construction BIM – general outline of tools and materials 31
    *2.6. Construction BIM – general outline of processes and
        procedures* 34
    2.7. Summary 36

PART II
Analysis: BIM/VDC form documents and guidelines,
and legal concepts   **41**

3 Standard of care and workmanlike performance   **43**

3.1. General concepts 44

3.2. BIM/VDC source documents – general industry research 48

    3.2.1 Center for Integrated Facility Engineering at
        Stanford University 48

    3.2.2 US Department of Commerce Cost Analysis of Inadequate
        Interoperability in the US Capital Facilities Industry 49

    3.2.3 Associated General Contractors Contractors' Guide to
        BIM – Edition 1 50

3.3. BIM/VDC source documents – BIM guidelines
and standards 52

    3.3.1 Indiana University BIM Guidelines & Standards for Architects,
        Engineers and Contractors 52

    3.3.2 State of Wisconsin Building Information Modeling (BIM)
        Guidelines and Standards for Architects and Engineers 54

    3.3.3 US Department of Veterans Affairs The VA BIM Guide 56
    State of Ohio Building Information Modeling Protocol 58

    3.3.4 New York City Department of Design and
        Construction BIM Guidelines 59

    3.3.5 National Building Information Modeling
        Standard – United States, v2 62

3.4. BIM/VDC source documents – form contracts 63

    3.4.1 AIA – E202™ (2008) Building Information Modeling
        Protocol Exhibit 64

    3.4.2 AIA – E203™ (2013) Building Information Modeling
        and Digital Data Exhibit 71

    3.4.3 AIA – G201™ (2013) Project Digital Data Protocol 78

    3.4.4 AIA – G202™ (2013) Project Building Information
        Modeling Protocol Form 79

    3.4.5 ConsensusDocs – 301™ Building Information
        Modeling (BIM) Addendum 83

3.5. Summary 92

4 Legal issues and claims considerations   **103**

4.1. Professional/contract – responsible control 104

4.2. Professional/contract – model development, use, and reliance 109

4.3. Professional/contract – legal status of the model 114

4.4. Technical – 2D–3D conversion 117

4.5. *Technical – interoperability* 120

4.6. *Technical – software version control* 122

4.7. *Technical – data loss, and data archiving and preservation* 123

4.8. *Technical – copyright and intellectual property* 124

4.9. *Summary* 126

5   **Methods and techniques for analysis of claims
    involving BIM/VDC**   131

5.1. *Reviewing contract documents for BIM and
    VDC responsibilities* 132

5.2. *Applicable standards of care and workmanlike performance* 133

5.3. *Analyzing the planned course of events* 134

5.4. *Investigating the actual course of BIM and VDC events* 135

5.5. *Comparing planned versus actual events, determining the
    impact, and explaining the results* 136

5.6. *Calculating BIM/VDC-specific damages* 138

5.7. *Scenario exploration* 138

6   **Preventative measures for enabling BIM/VDC success**   142

6.1. *Owners and facility managers* 142

6.2. *Architects and engineers* 142

6.3. *Contractors and general contractors* 143

6.4. *Subcontractors and fabricators* 143

6.5. *Drivers of change and impacts to claims involving
    BIM up to 2025* 144

Appendices   145
Index   205

Illustrations

Figures

1.1	Constructive change: altering planned vs. actual sequencing	6
1.2	Constructive acceleration: maintain original schedule	9
1.3	Delay – excusable *and* compensable	11
1.4	Disruption: labor inefficiencies	13
2.1	Design coordination process: general arrangement	29
2.2	Trade coordination process: general arrangement	35
3.1	BIM/VDC standard of care: analysis across knowledge bands	48

Tables

1.1	Typical construction claim types, forms, and damages	18
4.1	General legal issues in BIM/VDC form documents	105

Preface

This is not a book with an ideology.

That is, this book is not an attempt to prove or disprove that Building Information Modeling (BIM) or Virtual Design and Construction (VDC) will eradicate, reduce, or increase, the quantity, value, quality, or frequency of construction disputes, claims, or litigation.

This is a book that acknowledges the existence of two significant bodies of knowledge – technology and law – both with subsets concerned with designing and building. Part of the thesis here is that while each domain of knowledge has recently explored and analyzed BIM/VDC for their core constituents, a majority (not all) of popular literature on the topic as produced by either side has neither fully dissected nor explained their respective topics of interest by engaging the vocabulary of the other.

For example, while many in the BIM/VDC camp proclaim that the tools and process have the ability to reduce litigation, not all typically validate their position using vocabulary grounded in legal terms or construction claims concepts such as disruption, responsible charge, *Spearin*, privity, and so on. Similarly, while the legal camp might now include "BIM Execution Plan" as part of their vernacular, not all may be able to articulate the technical distinctions between an as-built construction BIM versus a conformed design BIM, nor why an owner might want either or both as deliverables.

Therefore, this book is intended to be a handbook towards bridging the divide between technology and law with respect to BIM/VDC. This will be done by utilizing the explicit vocabulary and deep bodies of knowledge of each domain for the collective benefit of both.

This book is intended as handbook for all stakeholders in the architecture, engineering, and construction (AEC) industry including owners, architects, mechanical and structural engineers, contractors, subcontractors, lawyers and students of design, construction, and law.

Executive summary

The AEC industry is incredibly complex, volatile and competitive. In a word, design, engineering, and construction are *risky*. Throw a dart at a map of the world and one will typically encounter a local AEC insurance and legal ecosystem with established rules of engagement for addressing issues of "mine" versus "yours" or perhaps better yet, "you owe me" versus "I don't owe you." To say that the AEC industry is accustomed to disputes, claims, and litigation would be a colossal understatement. Disputes are a fundamental element of the industry with a very long history. The *Code of Hammurabi*, a Babylonian law code dating from the eighteenth century BC, contains a series of provisions dealing specifically with construction disputes and includes some rather severe penalties for builders found to be negligent, namely death.[1] For the modern era, putting a precise dollar figure on the total value of global construction disputes would be a fool's errand but research suggests that, globally, recent annual amounts in controversy have collectively been worth billions of US dollars.[2]

Lest there be a tendency to moralize, even "good" projects might have claims and disputes. Change is a fact of life in design, engineering and construction projects. Inevitably, some of those changes may lead to disputes between parties. And while a claim won't necessarily lead to the need for mediation, arbitration or litigation, almost all contracts in the AEC industry are written to address the tendency towards change in the designing, engineering and building of things. For example, finding a contract for even the smallest of projects that does not include definitions and clauses regarding change orders, extensions of time, liquidated damages, and methods and procedures regarding dispute resolution would be the isolated exception as opposed to the rule. Legal jurisdictions around the world (both common law and civil law) have means and methods for articulating legal causes of action, interpreting contract clauses and the contemplation of monetary damages.

"But wait!" they say. The world is new and brave. BIM technologies and VDC processes, with their inherent transparency and collaborative approach, will enable clearer anticipation and better management of change in the first instance, thereby reducing the potential for claims leading to disputes and potential litigation. Perhaps. Perhaps not. While popular reports from early case studies appear to support the claims reduction thesis to a certain degree, evidence from other BIM/VDC-enabled projects reveal construction disputes to be alive and well.[3]

Will utilizing BIM-based spatial coordination (clashing) during the design and pre-construction phases of a project enable the prevention of many issues that might otherwise become the basis of a claim? Assuming proper BIM/VDC implementation and best practices in conjunction with existing best practices for design and construction in general, this *does* appear a reasonable bet. Can mobile and global positioning system (GPS)-enabled BIM/VDC technologies further capitalize on that pre-construction coordination work by enhancing and streamlining construction management and execution in the field? This also seems very likely. Would a dispute based on a subcontractor's approved substitution of a paint coating submittal that ultimately failed have been avoided with BIM/VDC technologies? The answer here appears less clear at the moment.

Determining cause and effect relationships between BIM/VDC and impacts on claims and litigation, and the outcome of those cases that go to trial, will also need to account for all manner of factors affecting the industry. For example, recent research suggests that while the average value of construction claims for the given sample set may have decreased, regional differences are not consistent, and that the overall time required to resolve those disputes has increased.[4] Further impacting the ability to investigate causation is the fact that the overall number of US trials is in decline.[5] In the United States this reduction of trials leads to a scarcity in reported cases which serve as the basis for common law. Without common law decisions, some have argued, "Construction law stops evolving."[6]

In light of all the above, the over-arching goal of this book is to provide a pragmatic exploration of the analysis of construction claims and disputes under the new lens of BIM/VDC. Is it possible that certain claims in the era of BIM/VDC will cease? Will others assume the clothes of traditional claims? Will new claim types evolve, and if so what will they look like? It is the investigation into questions such as these, not necessarily definitive answers, that this author considers critical for the AEC industry. This is because the use of BIM/VDC can profoundly and directly alter the production, delivery, and constitution of construction documents/drawings, quantity surveys, budgets/estimates, field-execution methods, and project schedules – each of which are central and critical components in the analysis of claims and disputes.

Accordingly, the primary objectives of this book are to: (1) present an overview of the typical construction claims environment in the United States, (2) investigate the continually evolving standards of care and workmanlike performance for design and construction professionals using BIM/VDC, and (3) provide industry professionals with a useful and practical tool and reference for effectively considering and analyzing construction claims and disputes on projects utilizing BIM technologies and VDC processes.

To accomplish these objectives this book is divided into two parts. Part I prepares the groundwork by examining traditional construction claims (Chapter 1). By providing a brief survey of the types of typical claims, the damages sought, and examples of typical associated contract clauses and/or case law, the reader is equipped with a baseline vocabulary and mindset. Next, a broad overview of the typical BIM/VDC tools, workflows, and processes of both designers and contractors is presented. This chapter introduces basic BIM/VDC concepts and prepares the reader to critically evaluate these new tools and processes against the legacy systems which are the environment in which traditional claims take root (Chapter 2).

Part II begins a synthesis of the claims synopsis and general outline of BIM/VDC practices from Part I to investigate notions of construction claims in an era of BIM/VDC. This process begins by evaluating what reasonably describes current baseline standard of care and workmanlike performance for BIM/VDC. This is accomplished through a review of primary (form contracts) and secondary (guidelines, etc.) BIM/VDC source documents (Chapter 3). Next, a categorization of BIM/VDC as it relates to legal concepts and possible claims issues is offered. Practically, the rubric for categorizing BIM/VDC claims is separated into two "sieves." The first sieve has been broadly termed "professional/contract" and leads off with a discussion of responsible control. The second sieve, "technical" explores issues such as two dimensional (2D)–three dimensional (3D) conversions, data loss, and so on. Needless to say, neither sieve is intended to block the passage or influence of concepts from one to the other (Chapter 4). Chapter 5 then presents typical methods and techniques for analysis of claims and considers their applicability for potential claims that either focus on, or contain elements of, BIM/VDC. Finally, Chapter 6 offers a series of pragmatic checklists to initiate risk mitigation and support successful implementation of BIM/VDC in the first instance.

The reader in search of absolutes will be forced to contend with the reality that this work, like any other, reflects a snapshot in time and that the BIM/VDC stream moves extremely fast for an industry not accustomed to speed when addressing anything new.

Notes

1 Johns, Claude Hermann Walter. *Babylonian and Assyrian Laws, Contracts, and Letters.* Vol. 6. The Lawbook Exchange, Ltd. (1904). Sections 229–233.
2 Zack, Jr., James G. *Trends in Construction Claims & Disputes.* Navigant Construction Forum. (December 2012). 5.
3 Post, Nadine M. *A Cautionary Digital Tale of Virtual Design and Construction: Insurance Settlement Related to a Building Information Model Shows That BIM without Communication Can Be Costly.* Engineering News Record (05/23/2011). Accessed on February 27, 2014. http://enr.construction.com/buildings/design/2011/0523-acautionarydigitaltale.asp.
4 EC Harris. *Global Construction Disputes 2012: Moving in the Right Direction.* (2012). Accessed on September 1, 2014. http://www.echarris.com/pdf/ech_global_construction_disputes_report_2012.pdf.
5 Galanter, Marc. "The Vanishing Trial: An Examination of Trials and Related Matters in Federal and State Courts." *Journal of Empirical Legal Studies* 1.3 (2004): 459–570.
6 Ness, Andrew D. *The Future of Construction Law and Claims.* 51st Annual Western Winter Workshop, AACE International, NV. (2012).

References

EC Harris. *Global Construction Disputes: Moving in the Right Direction.* (2012). Accessed on September 1, 2014. http://www.echarris.com/pdf/ech_global_construction_disputes_report_2012.pdf.

Galanter, Marc. "The Vanishing Trial: An Examination of Trials and Related Matters in Federal and State Courts." *Journal of Empirical Legal Studies* 1.3 (2004): 459–570.

Johns, Claude Hermann Walter. *Babylonian and Assyrian Laws, Contracts, and Letters.* Vol. 6. The Lawbook Exchange, Ltd. (1904). Sections 229–233.

Ness, Andrew D. *The Future of Construction Law and Claims.* 51st Annual Western Winter Workshop, AACE International. NV. (2012).

Post, Nadine M. *A Cautionary Digital Tale of Virtual Design and Construction: Insurance Settlement Related to a Building Information Model Shows That BIM without Communication Can Be Costly.* Engineering News Record (05/23/2011). Accessed on February 27, 2014. http://enr.construction.com/buildings/design/2011/0523-acautionarydigitaltale.asp.

Zack, Jr., James G. *Trends in Construction Claims & Disputes.* Navigant Construction Forum. (December 2012). 5.

Part I

Fundamentals

Construction claims, and BIM/VDC

1 Traditional construction claims

A review of existing literature concerning traditional construction claims and disputes reveals a path that is broad, clearly marked, and very well-trodden. Accordingly, the sections that immediately follow are intended to provide the design, engineering, construction, or construction law beginner with an introductory survey course in the broad topics of this established field. In simplistic terms and without a hint of cynicism, construction disputes are about money and why one party feels they are entitled to compensation from another party. The central issues of claims can vary widely from technical issues, to matters of law, to legal standards of care, and so on. As such, this chapter will briefly and succinctly convey the essence of typical construction claims and the monetary damages associated with each.

The goal of this chapter is to present a practical outline of a construction claims taxonomy that one might more easily recall when considering BIM and VDC issues. Whereas entire books might be (indeed have been) dedicated to individual claims topics, the intent here is to provide a general survey overview which is used as a basis for analysis in the chapters that follow. For example, while the concept of "delay" will be summarized, additional discussion of "concurrent delay" and the "apportionment of delay" will not be.

At the highest setting of the blade, this claims taxonomy will be divided into three primary segments: contractor claims against owners, owner claims against contractors, and claims in tort. The summaries within each segment will be drawn with single lines and will outline general shapes. Footnotes, the References and the List of cases will provide a gateway to further detail, nuance, and interpretation. Additionally, the summaries favor those aspects of claims which could find themselves first in line for renewed consideration under the lights of BIM and VDC. For example, defective and/or deficient construction documents versus differing site conditions.

Finally, in keeping with a survey methodology a brief note on general context and vocabulary: While non-traditional contracting methods including design–build (DB), integrated project delivery (IPD), and public private partnerships (P3s) continue to receive consideration by contracting parties (and in the case of P3s growing instances of legislative support[1] and implementation[2]), the typical construction claims analyzed here are rooted in the world of traditional design–bid–build (DBB) contracts. This is a contract framework wherein the owner has a contract with the architect or engineer to design the work, and a separate contract with a contractor to construct the design. Additionally, the general use of "contract" may include examples or discussion of private and/or government contracts, and/or any number contractual arrangements, including, for example, lump sum payment, guaranteed maximum price, and so on. Likewise, "design professional" may be used to refer to architects and/or all form of consulting engineers (mechanical, electrical, civil, etc.) with "contractor" similarly referring to a general contractor or construction manager of various type, unless specifically noted otherwise.

1.1 Contractor claims against owners

Claims on a construction project often arise in the contract for construction between the contractor and the owner. Thus, contractors and owners initially look to the four corners of the page in establishing the terms for the resolution of their differences. In doing so, it is important to appreciate that the page includes not only express clauses and provisions, but equally influential implied terms and industry custom. The well-crafted construction contract is no mere napkin scribble. As described by construction law scholar and practitioner James Arcet, construction contracts reflect multiple facets of the modern condition in regards to design and construction. Building codes, insurance, licensing and bonding, a long history of common law perspective on damages – each has a place.[3] Arcet also notes that construction contracts are influenced primarily by two types of knowledge, "construction knowledge" and "legal knowledge." The former privileges the holder in negotiating the requirements of the plans and specifications describing the work, the latter favors an understanding and anticipation of how contract terms have been interpreted by the courts.[4] While there may exist a current dearth of case law dealing specifically with BIM and VDC, the fact that these technologies and processes can fundamentally influence construction – both substantively with regards to means and methods, and procedurally in terms of typical duties, obligations or implied warranties – suggests that practical knowledge about both will well serve any party to a modern construction contract.

Contractors typically pursue claims against an owner under the following classifications: scope changes, acceleration claims, delay claims, and disruption claims. (Also included in this list are payment claims and termination claims, but for the summary purpose of this chapter a discussion of such claims is excluded.). These claims types often correspond loosely, if not directly, to given section headers or sub-headers in the written contract itself.[5] These classifications are not always mutually exclusive, hence a contractor's claim against an owner might include, for example, both a delay claim and a disruption claim.

As noted above, while claims are often pursued under specific contract clauses, there are also non-explicit, or implied, warranties in each construction contract including the contractor's workmanlike performance and the owner's implied warranty of contract documents. These doctrines establish, respectively, that the contractor has a duty to perform the work in accordance with the level of skill and care expected of the average qualified contractor in a given location, and that the owner will supply the contractor with drawings and specifications that are accurate, correct, and buildable. Implied warranties will receive further elaboration and consideration in later sections.

A brief summary of the typical construction claims brought by a contractor against an owner are presented below. These summaries will serve as the referential basis to more in-depth BIM and VDC analysis in subsequent chapters and sections.

1.1.1 Scope changes – overview

Defining *exactly* what work the contractor is going to do in fulfilling their obligations under the contract documents – the scope of work – is typically an extremely complex endeavor, some even suggesting impossible.[6] It usually requires the design professional to prepare detailed drawings (often voluminous) describing the quantitative aspects of the work, as well as detailed written specifications (often voluminous) describing the qualitative aspects of the work. Together, the drawings (graphic symbolism) and specifications (narrative text) set forth the requirements for the construction of the design.

The drawings and specifications must in turn be coordinated with the requirements of applicable building codes and industry standards. As anyone who has participated in a large design and construction project knows, it would not be unreasonable to imagine these documents filling a room, or even multiple rooms. In many cases the shear administrative task of keeping the specifications and drawings coordinated within and between themselves can be overwhelming.

It is from this proverbial mountain of information that the contractor must infer just what the expected end result is and how to get there on time and within budget. Hence, the colliding realities that serve as the nexus of claims. One variable includes extensive, highly detailed, interconnected and interdependent specifications, drawings, building codes and industry standards as prepared by the design professional. A second variable requires a contractor to interpret the aforementioned network of information. Both of which must contend with the inevitability of change on design and construction projects. Needless to say, these conditions provide a healthy environment for the birth of scope changes. Indeed, some consider a 5 to 10 percent cost growth due to scope changes within an acceptable normal range.[7]

In some instances scope changes are the result of explicit instructions by the owner. Such changes are referred to as *directed changes*. A simplistic example is, "Please add another floor to the parking garage that was not part of the original plan." Construction contracts often contain "changes" clauses that give the owner the flexibility to make such changes and formalizes the administrative processes of effecting the change. These changes clauses will typically define the type and timeliness of notice that must be given by the owner, and the amount of time the contractor has to respond with an estimate for the requested change. Typically, if the owner and contractor follow the process established by the changes clause, the owner achieves their desired outcome and the contractor receives an equitable adjustment to the contract cost and/or schedule. If, however, the owner and contractor cannot agree on the exact scope of work for the directed change and the related costs for the change, a claim may ensue.

In contrast with a directed change is the concept of a constructive change. In a constructive change claim a contractor and an owner may have a disagreement over what the original contract documents actually show, say, or mean. Or, there may be events or conditions such as improper owner rejection of work, or excessive testing requirements that may result in a scope change. Thus, a constructive change is conduct (either action or inaction) by an owner that is not a formal change order (a directed change) but which has the effect of causing the contractor to perform the work in a manner different from what the terms of the original contract required. Furthermore, a constructive change does not necessarily mean there is a change to the character and quantity of the work.[8]

In contrast to the previous example of an owner directed change "add another parking level," a constructive change might take the form of an owner action (or inaction) around sequencing of work. For example, consider a construction drawing showing elements A and B. The contractor reviews the drawings and, bringing to bear their expertise in the means and methods of construction, prepare a bid to complete the scope of work for installing A and B. Their bid includes plans to install A before B. The owner accepts the bid and the contractor begins the work. However, subsequent actions or inactions on the part of the owner result in the contractor being forced to install B before A. The effect of forcing the contractor to alter his planned sequence or duration for installing A before B may cause the contractor to experience cost and/or schedule inefficiencies without altering the scope of work originally defined by the contract and used as the basis of the bid, that is, installing A and B (Figure 1.1).[9]

Generally accepted theories by which a contractor might make a claim for a constructive scope change include, amongst others, defective and deficient contract documents, and implied warranties and duties. In the interest of a longer term focus on BIM and VDC issues, these two items are summarily reviewed here.

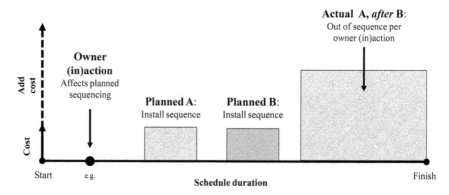

Figure 1.1 Constructive change: altering planned vs. actual sequencing

It is generally accepted that defects and deficiencies in the contract documents present a target-rich environment for construction claims.[10] While no set of contract documents is ever imagined as perfect or error free, the liability for errors and omissions lies with the preparer.[11] Furthermore, an owner who supplies drawings and specifications to a contractor implies a warranty to the contractor that the plans are suitable for construction and free from defect.[12] This latter point, commonly known as the *Spearin* doctrine is a fundamental tenet of US construction law and is based on a United States Supreme Court decision almost 100 years old at the publication of this book. *United States* v. *Spearin*[13] involved a contract to build a dry dock at the Brooklyn Navy Yard in New York. A defect in the plans and specifications lead to an event compromising the dry dock excavation. Unable to resolve the contractor's claim that the Government as owner for the plans and specifications needed to take responsibility for the event, the Secretary of the Navy annulled the contract, and the contract was ultimately completed by a different contractor. In their decision, the *Spearin* court held that Spearin, bound by contract to build according to plans and specifications provided by the Government, was not responsible for defects in those plans and specifications and that there is an implied owner warranty that if the plans and specifications are followed, the resulting construction will be satisfactory. The contractor's duty to check the plans did not impose on them the obligation to determine their adequacy to accomplish the purpose. The *Spearin* decision embodies a "line in the sand" separating design and construction in traditional design–bid–build contracts. Again, while approaching 100 years old, *Spearin* is still regularly cited in US construction law.

The implied warranty of "workmanlike construction" establishes that a contractor has a duty to perform the work in accordance with the level of skill and care expected of the average qualified contractor in a given location. For example, the Tennessee Supreme Court in *Federal Ins. Co.* v. *Winters*[14] determined that a roofing contractor, who had hired another independent roofing contractor to fix a leak in his original work which subsequently caused a fire at the structure, nevertheless had an implied, non-delegable duty to install the roof in a careful, and workmanlike manner. As part of its opinion, the *Federal* court cited *Bowling* v. *Jones*[15]:

> Once a builder undertakes a construction contract, the common law imposes upon him or her a duty to perform the work in a workmanlike manner, and there is an implied agreement that the building or work performed will be sufficient for the particular purpose desired or to accomplish a certain result. Thus, failure to perform a building contract in a workmanlike manner constitutes a breach of the contract.

1.1.2 Acceleration – overview

Acceleration claims are claims based on time. And as the age-old refrain goes, "time is money." Acceleration, as can be logically inferred, involves speeding up the work. The result of speeding up the work is that the contractor is forced to complete the original scope of work in less time than originally planned or budgeted for. This causes the contractor to, typically, increase his manpower to meet the time constraint either through adding overtime or enlisting additional labor forces, or sometimes both. Thus, with an acceleration claim the contractor is looking to recoup any extra costs associated with additional labor costs and/or manpower required as a result of speeding up the work.

As with scope changes, acceleration can be segregated along lines of *directed acceleration* or *constructive acceleration*.[16] Directed acceleration takes the form of explicit owner direction – "Mr. Contractor you must now complete the parking garage by March 15th of this year instead of October 31st." Many construction contracts contain an acceleration clause, enabling the owner to direct such acceleration and requiring the owner to compensate the contractor for the acceleration costs. A generic example of a contract clause concerning acceleration might read, in part:

> The Owner may, at any time, by written Request for Proposal designated or indicated to be a change order, make changes in the Work, within the general scope of the Contract, including changes (i) in the specifications (including drawings and designs); or (ii) directing acceleration in the performance of the Work.

An acceleration clause such as this would enable the owner from the preceding example to request, in writing, a new completion date of March 15th and give the contractor the process and requirements – likely defined elsewhere in the contract – for pricing the additional manpower and/or labor costs required to meet the accelerated completion date. If all parties adhere to the terms of the contract, the directed acceleration may cause no further issue. An acceleration request is made, a solution is proposed, priced, and accepted, and the work is completed by the new date.

However, the complexity of design and construction projects often present less clear-cut scenarios. Enter a contractor's claim for constructive acceleration.[17] With constructive acceleration there is no formal directive from the owner to complete the scope of work in a shorter amount of time. However, a series of events accumulate to effect the same result, hence the acceleration is inferred, or constructed. That series of events initiates with the contractor notifying the owner that an *excusable delay* has occurred and requesting a time extension to the schedule to account for the excusable delay. The owner, however, then refuses to grant the extension within a reasonable time frame. The refusal can be either explicit or implied through inaction. The contractor is then forced to maintain the original schedule, in reality – speed up. The sum effect is that the contractor must complete the original scope of work, plus additional time added by the excusable delay, within the originally scheduled time frame (Figure 1.2).

Definitions of what constitutes an excusable delay are typically defined as something beyond the control of, and not caused by the fault or negligence

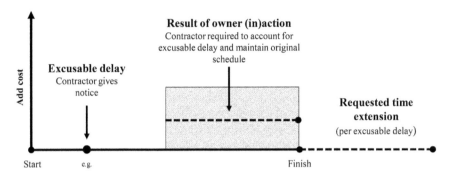

Figure 1.2 Constructive acceleration: maintain original schedule

of, the contractor. Specific examples are usually stated in the construction contract and may include: defective plans and specifications or owner-ordered changes. For example, Article 6.3.1 of the ConsensusDocs 200™ *Standard Agreement and General Conditions Between Owner and Constructor (Lump Sum Price)* reads, in part:

> Examples of causes beyond the control of the Constructor include, but are not limited to, the following: (a) acts or omissions of the Owner, the Design Professional or Others; (b) changes in the Work or the sequencing of the Work ordered by the Owner, or arising from decisions of the Owner that impact the time of performance of the Work; (c) encountering Hazardous Materials, or concealed or unknown conditions; (d) delay authorized by the Owner pending dispute resolution or suspension by the owner . . . (e) transportation delays not reasonably foreseeable; (f) labor disputes not involving the Constructor; (g) general labor disputes impacting the Project but not specifically related to the Worksite; (h) fire; (i) terrorism; (j) epidemics; (k) adverse governmental actions; (l) unavoidable accidents or circumstances; (m) adverse weather conditions not reasonably anticipated . . .

1.1.3 Delay – overview

Delay claims, like acceleration claims, are fundamentally about time. As the name implies, delay claims involve increases to the amount of time required to complete the scope of work beyond what was contemplated when the contract was executed. Delay can cut both ways. A contractor may make a claim for delay against the owner – "I only budgeted for X number of days, and this job took Y." Vice versa, the owner may claim delay against the contractor – "I planned on being open for business on Jan. 1 and now I cannot occupy until April 1." This section discusses a contractor's claim for delay against an owner.

In comparison to a contractor's acceleration claim which looks to recover additional costs incurred with adding overtime hours and/or bringing in more manpower to meet a compressed schedule, with a delay claim the contractor seeks to be compensated for time-related items such as increased home- and field-office overhead, idle construction equipment, or additional costs of financing.[18] In common parlance, the contractor's delay claim is looking to recover not only increased material and/or labor costs, but the "costs of doing business" brought about by the increased amount of time spent on the job. Of particular significance in delay claim damage calculations are calculations for home-office overhead – the continuous administrative and executive costs of a contractor – as opposed to operating charges specific to the project that has been delayed. Calculating the home-office overhead portion of damages in a delay claim is typically computed using the *Eichleay* formula. The *Eichleay* formula derives its name from the 1960's case *Appeal of Eichleay Corp.*, ASBCA No. 5183.[19] The *Eichleay* formula assists in determining an estimate of the portion of home-office overhead that should be allocated to a particular job because of the increased duration of the project. While typically utilized in government delay cases, the case law on *Eichleay* interpretations for other jurisdictions is varied.[20]

A contractor's delay claim will be divided along the lines of *excusable* or *non-excusable* delays. At the most basic level, an excusable delay is one that the contractor is not contractually responsible for. By definition, the contractor is *excused* from the liability of the delay and is entitled to an extension of time in the contract to complete the work (and perhaps financial compensation as will be discussed below). Typical examples of excusable delays include: changes, labor disputes, fire, abnormal weather conditions, or an act of neglect of the owner, architect or other party contracted by the owner. A *non-excusable* delay, as the name implies, is a delay where it is determined that the contractor is responsible. Therefore, the contractor is not excused from liability and, depending on circumstances, may be required to pay damages compensation to the owner.

At a second tier of classification, excusable delays are further considered in terms of being either *compensable* or *non-compensable*. As can be intuitively inferred, non-compensable delays are delays that entitle the contractor to a time extension to complete the scope of work, but no additional financial compensation. General Condition 8.3.1 of *AIA Document A-201 (2007)* is an example of language providing for a time extension on an excusable delay:

> If the Contractor is delayed at any time in the progress of the Work by an act or neglect of the Owner or the Architect, or by any employee of either, or by any separate contractor employed by the Owner; or by changes ordered in the Work; or by labor disputes, fire, unusual delay in deliveries, unavoidable casualties, or any causes beyond the Contractor's control; or by delay authorized by the Owner pending mediation and arbitration; or by other causes that the Architect determines may justify the delay, then the Contract Time shall be extended by Change Order for such reasonable time as the Architect may determine.

With a compensable delay the contractor seeks not only a time extension, but monetary damages as well (Figure 1.3). Many owners to private contacts suggest or require so-called "no damages for delay" clauses to be included in the contract that preclude the contractor from obtaining any such damages. In the United States, the fifty states have various interpretations to the enforceability of such clauses.[21] However, judicial decisions have regularly held that a "no damages" clause notwithstanding, contractors are entitled to recover damages when delays are caused by acts of the owner. Furthermore, the contractor's right to recover damages for owner-caused delay are implied in each construction contract.[22] For example, in *Northeast Clackamas County Electric Cooperative, Inc.* v. *Continental Casualty Company*,[23] the court stated in relevant part:

> In every express contract for the erection of a building or for the performance of other constructive work, there is an implied term that the owner, or other person for whom the work is contracted to be done, will not obstruct, hinder, or delay the contractor, but on the contrary, will always facilitate the performance of the work to be done by him.

1.1.4 Disruption – overview

Whereas acceleration and delay claims are focused on the effects of shortening or lengthening of the time duration of the schedule, disruption claims are concerned with events. Disruption claims involve a contractor's assertion that certain events – disruptions – precluded them from executing and completing the scope of work in the manner they contemplated and originally bid for the work. Examples of disruptions that a contractor might experience include: incomplete drawings, scope changes, or failure to respond to requests for information. In turn, each disruption might result in a labor inefficiency, including stacking of trades, re-sequencing, overtime, or field installations versus pre-fabrication. In contrast to delay claims, a disruption claim may not necessarily extend the amount of time the contractor spends on the job.

An essential difference between delay claims and disruption claims is the type of monetary damages sought. With delay claims, the damages sought often

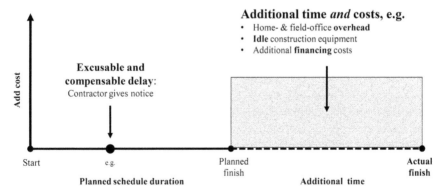

Figure 1.3 Delay – excusable *and* compensable

include costs for extended overhead for the contractor's field and home offices as a result of a prolonged schedule. In a disruption claim the typical damages include costs for additional labor, increased material and equipment costs, and, significantly, the increased costs for field labor inefficiency caused by altered working conditions or overtime.[24] To this last point, some authors have noted that being able to prove lost productivity to a trier of fact is among the most difficult tasks in construction claims.[25] For the purpose of the research here, it is worthwhile to note that Mechanical Contractors Association of America (MCAA) *Change Orders, Productivity, Overtime: A Primer for the Construction Industry (2012)*[26] specifically identifies the BIM process as another type of labor that can be impacted by project events and that such costs should be carefully reviewed in the preparation of delay and or disruption claims (Figure 1.4). Another distinction between delay and disruption claims involves so-called "no damages for delay" clauses which, as previously discussed, could preclude monetary damages for delays, but traditionally do not bar the recovery of damages for disruptions.[27]

1.2 Owner claims against contractors

Examples of owner claims against the contractor include delay claims, property damage, and performance failures. Similar to a contractor's claim for delay damage, an owner's delay claim would seek to recover damage done to its business position. The types of damages for the late completion of a project might include claims for lost rental income, interest on project loan financing, real estate taxes, or perhaps even lost profits on missed sales. Some contracts will seek to limit these types of damages. For example the AIA *B101-2007 Standard Form Agreement Between Owner and Architect* includes a clause requiring the owner and architect to mutually waive consequential damage claims against each other.[28]

Construction contracts might include provisions for liquidated damages. Liquidated damages are damage amounts that the owner and contractor negotiate when the contract is being formed and reflect their mutual agreement as to the daily value of delay by the contractor. Simply put, if the liquidated damages are $100/day and the project delay is ten days in duration, the owner would receive liquidated damages of $1,000.00.

1.3 Torts

Having now spent some energy considering how various types of claims are pursued under the terms of a contract, the contract must now be temporarily set aside, or at least forced to share the spotlight. This is because the possibility exists that a construction project might also see claims brought under the law of torts. A tort, broadly defined, is where a legal wrong has been committed upon a person or property, but there is no contract between the parties. Negligence – the concept of failing to exhibit the care which a reasonably prudent person would use under similar circumstances – is a type of tort that most people are familiar with. One may not have a contract with

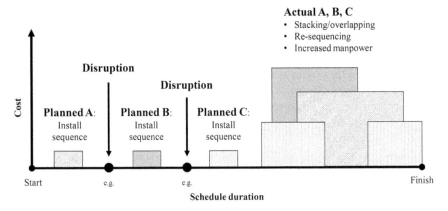

Figure 1.4 Disruption: labor inefficiencies

the man he passes on the street, but there is the expectation that both parties will act in a way that seeks to avoid harming the other.

Reviewing the general literature, treatises, and case law of torts leaves a reader confident in one regard – the topic is multifaceted and defies tidy explanation. Even the recognized legal authority on torts concedes the difficulty in perfectly defining a tort, or distilling a singular guiding principal as to when compensation should be paid out.[29] One should be prepared for nuance and interpretation. And to be sure, nuance thrives in the arena of design and construction where there is typically multiple project participants engaged in a common pursuit, but where many of the relationships between stakeholders are not based in contract.

A survey of tort cases in design and construction can lead down disparate paths. Is a supplier of concrete liable to an owner for the future deterioration of concrete slabs and foundations where an incorrect concrete mixture contained submicroscopic damage even though there was no present damage to the slabs and foundation? In *Mesa Vista* v. *Cal. Portland Cement Co.*, a California appeals court held, "yes."[30] Can an architect be held liable for the suicide of a prisoner committed in a prison designed by said architect? US courts have typically said, "no."[31] Can a contractor be held liable for failure to install a proper hot water control valve which caused a paraplegic plaintiff to sustain second and third degree burns where the plaintiff twice failed to test the water temperature before placing their feet directly in the bathtub? In *Tirella* v. *Am. Props. Team, Inc.*, the New York State Supreme Court affirmed, "yes."[32] Cases such as these (and innumerable others) offer ample grounds for discussion and debate. However, for the given purpose of analysis related to BIM and VDC, only tort cases that have addressed the liability of architects and engineers to contractors with whom they are not in contract are singled out here.

As discussed in earlier sections, construction claims typically entail one party alleging breach of contract against the other. The connection or relationship established between the parties to a contract is known as "privity of contract."

Thus, generally speaking, there is typically no privity of contract between designers and contractors. Owners typically hold a contract with the architect to design a project and a separate contract with the contractor to construct the design. Even assuming an owner's implied warranty to the contractor of the completeness of the architect's contract documents, there is no contractual relationship created between the architect and contractor because they've not directly entered into any agreement. Some have suggested this framework, as perpetuated by form contracts, creates an incomplete triangle.[33] Accordingly, a contractor seeking damages against an architect with whom he is not in privity might pursue a negligence claim in tort.

Absent a contract, a contractor pursing such a negligence claim would encounter the economic loss "rule." If a loss (financial damage) is economic, it is a loss that is not caused by physical injury or property damage. The financial damage done is not caused by someone having been, say, maimed, nor there having been actual damage to a structure. The loss is exclusively economic, for example lost income, and therefore typically not recoverable in a negligence claim. As the court in *Indemnity Ins. Co. v. American Aviation* succinctly captured in their footnotes, "Economic losses are, simply put, disappointed economic expectations."[34] Economic *expectations* are the provenance of contracts where the parties would contemplate those types of actual losses, for example contract clauses regarding liquidated damages due an owner for project delays where the damages are not penalties, but rather the parties' negotiated estimate or expectations of the actual value of a delay.[35] Conversely, the law of torts is concerned with affording compensation for physical injury and/or property damages sustained by one person as a result of the conduct of another, absent a contract.[36] Accordingly, the definition of the economic loss rule offered by the *Indemnity* court is helpful in its summarizing clarity, "The . . . rule is a judicially created doctrine that sets forth the circumstances under which a tort action is prohibited if the only damages suffered are economic losses."[37] If the contractor is not physically injured, or property is not physically damaged, and the only damages sought are economic – then the contractor has no actionable claim in tort against the architect.

As one might expect, the economic loss "rule" has seen exceptions when contemplating the lack of contractual privity between architects and contractors. Exceptions have included: foreseeability, special relationship, and negligent misrepresentation.[38] For example, the State of Florida has provided an exception on the grounds that in the course of rendering professional services, architects should anticipate that the information they are providing to contractors will be used by the contractor for estimating construction costs, thereby creating a special relationship between an architect and contractor who are not in direct contract.[39] In reaching their decision in *Hewett-Kier Const., Inc. v. Lemuel Ramos and Associates, Inc.*,[40] the District Court of Appeals of Florida, 4th District relied upon both Florida case law supporting negligence actions against professionals despite a lack of direct contracts,[41] and Section 552 of the American Law Institute's treatise, *Restatement (Second) of Torts*.[42] Section 552, *Information Negligently Supplied for the Guidance of Others* reads in relevant part:

(1) One who, in the course of his business, profession or employment, or in any other transaction in which he has a pecuniary interest, supplies false information for the guidance of others in their business transactions, is subject to liability for pecuniary loss caused to them by their justifiable reliance upon the information, if he fails to exercise reasonable care or competence in obtaining or communicating the information.

(2) Except as stated in Subsection (3), the liability stated in Subsection (1) is limited to loss suffered

(a) by the person or one of a limited group of persons for whose benefit and guidance he intends to supply the information or knows that the recipient intends to supply it; and

(b) through reliance upon it in a transaction that he intends the information to influence or knows that the recipient so intends or in a substantially similar transaction.

Similarly, the Supreme Court of Pennsylvania has found that a building contractor, who was not in privity of contract with an architect, may nonetheless maintain a negligent misrepresentation claim against the architect for alleged misrepresentations in the architect's plans and specifications which caused the contractor to suffer purely economic damages. In *Bilt-Rite Contractors, Inc. v. Architectural Studio*[43] the architect, under contract with a public school district, prepared plans and specifications for bidding purposes. The bid drawings included the installation of an aluminum curtain wall system, sloped glazing and support mechanisms. Further, the architect expressly indicated in the bid documents that those elements could be constructed employing normal construction means and methods. The school district subsequently included the architect's plans and specifications in its solicitations for bids, and the lowest responsible bidder was awarded the contract for construction. Once construction began, the contractor discovered that the aforementioned curtain wall, glazing, and support systems could not be installed utilizing traditional means and methods, but required special techniques which increased the cost of the work. The contractor alleged negligent misrepresentation under Section 552 of the *Restatement (Second) of Torts* claiming the architect's specifications with regard to the systems were false and/or misleading.

The court noted that in support of their original complaint, the contractor specifically cited Comment (h) of Section 552, and Illustration (9) thereof, as exactly describing their claim:[44]

(h) *Persons for whose guidance the information is supplied.* The rule stated in this Section subjects the negligent supplier of misinformation to liability only to those persons for whose benefit and guidance it is supplied. In this particular his liability is somewhat more narrowly restricted than that of the maker of a fraudulent representation (see §531), which extends to any person whom the maker of the representation has reason to expect to act in reliance upon it.

(9) The City of A is about to ask for bids for work on a sewer tunnel. It hires B Company, a firm of engineers, to make boring tests and provide a report showing the rock and soil conditions to be encountered. It notifies B Company that the report will be made available to bidders as a basis for their bids and that it is expected to be used by the successful bidder in doing the work. Without knowing the identity of any of the contractors bidding on the work, B Company negligently prepares and delivers to the City an inaccurate report, containing false and misleading information. On the basis of the report C makes a successful bid, and also on the basis of the report D, a subcontractor, contracts with C to do a part of the work. By reason of the inaccuracy of the report, C and D suffer pecuniary loss in performing their contracts. B Company is subject to liability to C and to D.

In an extensive opinion, the *Bilt-Rite* court found that case law from Pennsylvania, as well as other jurisdictions, supported a decision to apply liability as constructed in Section 552 to the architect/contractor relationship in an allegation of negligent misrepresentation. Amongst others cases, the court quoted from the Arizona Supreme Court decision in *Donnelly Const. Co.* v. *Oberg/Hunt/Gilleland*,[45] which stated, in part:

Design professionals have a duty to use ordinary skill, care and diligence in rendering their professional services . . . When they are called upon to provide plans and specifications for a particular job, they must use their skill and care to provide plans and specifications which are sufficient and adequate . . . This duty extends to those with whom the design professional is in privity . . . and to those with whom he or she is not.

The *Bilt-Rite* court ruled that there was no privity of contract requirement, and that the economic loss rule did not bar the contractor from recovery of economic damages. The court did recognize that an architect's liability is not limitless, but felt that Section 552 was adequate in defining those limits. Accordingly, the court could find no reason why Section 552 should not apply to architects and that the architect's contractual duties to the party it was contracted to flowed on to foreseeable third parties, that is, contractors.[46]

Privity of contract, the economic loss doctrine and its exceptions, duties owed to foreseeable third parties – each garnered attention with the emergence of BIM/VDC. One of the earliest objections to BIM usage was concern that highly collaborative BIM tools and VDC processes would create privity of contract between design and construction parties where none had previously existed.[47] To address this concern early industry form contract documents such as the ConsensusDocs 301 *BIM Addendum* includes language stating that the use of the *Addendum* is not intended to create privity of contract other than as exists in law or as per any governing contracts to which the *Addendum* is attached.[48] Likewise, the AIA E203 (2013) *Building Information Modeling and Digital Data Exhibit* states that it does not create any third-party beneficiary rights other than those specifically created within the Exhibit.[49]

Acknowledging the early industry need for such language, the constant progress of technology demands continuously asking what, "reasonable care" and/or "ordinary skill" such as utilized in the *Restatement (Second) of Torts*, and *Donnelly* decision means with respect to BIM and VDC. For example, during the course of writing this book (June 2014), the US National Council of Architectural Registration Boards posted commentary related to proposed changes to their Intern Development Program for training and preparing architects for licensure:

> Technology has drastically increased the scope and speed with which intern architects design and document today's projects. Current digital tools require interns to learn how a building is put together much earlier in the design process. Furthermore, many aspects of design documentation that used to require hours of labor can now be completed in minutes. One hour in a firm today is much different than one hour in a firm when IDP was first created.

Even conceding the above as commentary, and not yet established policy, given the source, "ordinary" and "reasonable" professional design services would appear to now include BIM and VDC insofar as architectural interns are very often the ones preparing design and construction documentation under the responsible charge of licensed architects. Likewise, a contractor's workmanlike performance and duty to inquire with respect to patent ambiguities in bid documents might be enhanced by BIM and VDC.[50] Retrospectively applied to the discussion above, examples might include time-sequenced visualizations of the means and methods of a curtain wall system installation as was part of the alleged negligent misrepresentation in *Bilt-Rite*.

Given the multiple interpretations of the courts in deciding cases of negligent liability of the designer there is no reason to anticipate absolute resolution any time soon. However, to suggest that the transparent nature of BIM tools and VDC processes may result in the courts casting additional and new interpretations into the dynamic stream of construction torts seems fair.

1.4 Summary

This chapter presented a practical outline of typical claims concepts including scope change, acceleration, delay, disruption, and claims in tort (Table 1.1). The literature and case law of construction claims is expansive and additional reading is required to fully appreciate the depth and breadth of construction claims. This summary of general terms and sources such as the *Spearin* doctrine, implied warranty of workmanlike manner, constructive acceleration, privity of contract,

Table 1.1 Typical construction claim types, forms, and damages

Claim type (examples)	Forms (examples)	Damages (examples)
Acceleration	• Directed • Constructive	• Increased hours • Increased manpower
Delay	• Excusable • Non-excusable	• Home/field-office overhead • Cost of financing
Disruption		• Additional manpower • Labor inefficiency
Tort	• Negligence • Negligent misrepresentation	• Economic

Restatement (Second) of Torts, and so on offers a baseline of traditional construction claims themes and vocabulary for critically engaging BIM/VDC in the following chapters.

Notes

1 State of Maryland HB560 (Chapter 5). Accessed on February 22, 2014: http://mgaleg. maryland.gov/2013RS/Chapters_noln/CH_5_hb0560e.pdf.
2 Ohio DOT. Accessed on February 22, 2014: http://www.dot.state.oh.us/Divisions/ InnovativeDelivery/Pages/Projects.aspx.
3 Acret, James. *National Construction Law Manual.* 4 ed. BNi Publications, Inc. (2002). 103.
4 Acret. 103.
5 See, for example, American Institute of Architects. *A201 (2007) General Conditions of the Contract for Construction* (2007); Singapore Building and Construction Authority. *Public Sector Standard Conditions of Contract for Construction Works 2008.* 6 ed. (2008).
6 See, for example, Sweet, Justin. *Sweet on Construction Industry Contracts: Major AIA Documents.* New York: Wiley (1996). 19.
7 *AACE International Recommended Practice No. 25R3* (2004). Citing Schwartzkop "Estimating Lost Labor Productivity in Construction Claims." 4.
8 For example, Hatem, D. J., & Kalayjian, W. H. "Change in Scope Claims." In R. F. Cushman (Ed.) *Proving and Pricing Construction Claims.* 3rd ed. Aspen Publishers (2001). 181–196.
9 Ibid.
10 For example, Haese, G. H., & Dragelin, T. J. "Types of Claims." In R. F. Cushman (Ed.) *Proving and Pricing Construction Claims.* 3rd ed. Aspen Publishers (2001). 1–66.
11 Hatem and Kalayjian. 187.
12 Acret. 95.
13 248 US 132 (1918).
14 354 S.W.3d 287 (Tenn. 2011).
15 300 S.W.3d 288, 291 (Tenn.Ct.App.2008).
16 For example, Hurlbut, S. B., Santiago, S. J., & Pocalyko, P. W. "Acceleration Claims." In R. F. Cushman (Ed.) *Proving and Pricing Construction Claims.* 3rd ed. Aspen Publishers (2001). 97–136.

17 Certain research notes that the term "constructive acceleration" is a uniquely American creation and that while other international jurisdictions recognize a claim rising in similar fact patterns and fulfilling similar typical requirements, none of those jurisdictions uses the term. See, for example, Zack Jr., James G. "The Basics of Constructive Acceleration." *Cost Engineering* 54.1 (2012): 4.

18 For example, Ficken, B. W., & Fertitta, T. D. "Delay Claims." In R. F. Cushman (Ed.) *Proving and Pricing Construction Claims*. 3rd ed. Aspen Publishers (2001). 29–66.

19 60-2 B.C.A. (CCH) ¶ 2688 (1960).

20 Pocalyko, P. W., Hooper, I., Libor, M., & Peter, M. "Using the *Eichleay* Formula for Home Office Overhead Claims." In R. F. Cushman (Ed.) *Proving and Pricing Construction Claims*. 3rd ed. Aspen Publishers (2001). 353–378.

21 For example, Dodd, Michael and Findlay, Duncan J. *State-by-State Guide to Design and Construction Contracts and Claims*.2nd ed. Aspen Publishers (2012).

22 Gavin, D. G., Toomey, D. E., & Guernier. W. D. "Disruption Claims." In R. F. Cushman (Ed.) *Proving and Pricing Construction Claims*. 3rd ed. Aspen Publishers (1996). 67–96.

23 221 F.2d 329 (9th Cir. 1955).

24 Gavin, et al.

25 *AACE International Recommended Practice No. 25R3* (2004).

26 Mechanical Contractors Association of America. *Management Methods Bulletin C02-2011*. MCAA (2012).

27 Gavin, et al.

28 §8.1.3.

29 Prosser, W. L., & Keeton, P. *Prosser and Keeton on Torts*. West Publishing (1984). 1.

30 12 Cal. Rptr. 3d 863 (Ct. App. 2004).

31 For example, *Estate of Stephen Puza* v. *Carbon County*, 586 F. Supp. 2d 271 (M.D. Pa. 2007).

32 145 A.D.2d 724, 535 N.Y.S.2d 252 (App. Div. 1988).

33 For example, Sweet. 32.

34 So. 2d 532 (Fla. 2004). Note 1.

35 Arcet. 2.

36 Prosser. 1.

37 *Indemnity Ins. Co.* v. *American Aviation*.

38 Caine, C. P., & Thomas, H. R. "Negligent Tort Liability of the Design Professional." *Journal of Legal Affairs and Dispute Resolution in Engineering and Construction* 5.1 (2012): 45–52.

39 Arcet, citing *Hewett-Kier Const., Inc.* v. *Lemuel Ramos and Associates, Inc.*, 775 So. 2d 373 (Fla. Dist. Ct. App. 2000). 176.

40 775 So. 2d 373 (Fla. Dist. Ct. App. 2000).

41 *Moransais* v. *Heathman*, 744 So. 2d 973 (Fla. 1999).

42 *Hewett*. Note 1. (Also, while not of direct pertinence here, engaging opinion and cultural context on the *Restatements* as a major law reform project in the US can be found in Gant Gilmore's, *The Ages of American Law*. Yale University Press, 1977. 3.)

43 866 A.2d 270, 581 Pa. 454 (2005).

44 Ibid. 275.

45 677 P.2d 1292, 139 Ariz. 184 (1984).

46 Ibid. 286.

47 For example, Larson, D. A., & Golden, K. A. "Entering the Brave, New World: An Introduction to Contracting for Building Information Modeling" *William. Mitchell Law Review* (2007). 34, 75.

48 ConsensusDocs 301, ¶1.2.

49 AIA E203 (2013) §1.2.1.

50 *Infra* 3.4.1.1.

References

AACE. *International Recommended Practice No. 25R3* (2004). Citing Schwartzkop "Estimating Lost Labor Productivity in Construction Claims."

Acret, James. *National Construction Law Manual.* 4th ed. BNi Publications, Inc. (2002).

American Law Institute. *Restatement (Second) of Torts: Information Negligently Supplied for the Guidance of Others.* §552 (1977).

Caine, C. P., & Thomas, H. R. "Negligent Tort Liability of the Design Professional." *Journal of Legal Affairs and Dispute Resolution in Engineering and Construction* 5.1 (2012): 45–52.

Dodd, Michael and Findlay, J. Duncan. *State-by-State Guide to Design and Construction Contracts and Claims.* 2nd ed. Aspen Publishers (2012).

Ficken, B. W., & Fertitta, T. D. "Delay Claims." In R. F. Cushman (Ed.) *Proving and Pricing Construction Claims.* 3rd ed. Aspen Publishers (2001).

Gavin, D. G., Toomey, D. E., & Guernier. W. D. "Disruption Claims." In R. F. Cushman (Ed.) *Proving and Pricing Construction Claims.* 3rd ed. Aspen Publishers (1996).

Gilmore, Grant. *The Ages of American Law.* Yale University Press (1977).

Haese, H. G., & Dragelin, T. J. "Types of Claims." In R. F. Cushman (Ed.) *Proving and Pricing Construction Claims.* 3rd ed. Aspen Publishers (2001).

Hatem, D. J., & Kalayjian, W. H. "Change in Scope Claims." In R. F. Cushman (Ed.) *Proving and Pricing Construction Claims.* 3rd ed. Aspen Publishers (2001).

Hurlbut, S. B., Santiago, S. J., & Pocalyko, P. W. "Acceleration Claims." In R. F. Cushman (Ed.) *Proving and Pricing Construction Claims.* 3rd ed. Aspen Publishers (2001).

Larson, D. A., & Golden, K. A. "Entering the Brave, New World: An Introduction to Contracting for Building Information Modeling." *William Mitchell Law Review* (2007): 34, 75.

Maryland. HB560 (Chapter 5). Accessed on February 22, 2014. http://mgaleg.maryland.gov/2013RS/Chapters_noln/CH_5_hb0560e.pdf.

Mechanical Contractors Association of America. *Management Methods Bulletin C02-2011.* MCAA (2012).

Ohio. Department of Transportation. Accessed on February 22, 2014. http://www.dot.state.oh.us/Divisions/InnovativeDelivery/Pages/Projects.aspx.

Pocalyko, P. W., Hooper, I., Libor, M., & Peter, M. "Using the *Eichleay* Formula for Home Office Overhead Claims." In R. F. Cushman (Ed.) *Proving and Pricing Construction Claims.* 3rd ed. Aspen Publishers (2001).

Prosser, W. L., & Keeton, P. *Prosser and Keeton on Torts.* West Publishing (1984). 1.

Sweet, Justin. *Sweet on Construction Industry Contracts: Major AIA Documents.* New York: Wiley (1996).

Zack Jr., J. G. "The Basics of Constructive Acceleration." *Cost Engineering* 54.1 (2012): 4.

List of cases

Bilt-Rite Contractors, Inc. v. Architectural Studio, 866 A.2d 270, 581 Pa. 454 (2005)

Bowling v. Jones, 300 S.W.3d 288 (Tenn. Ct. App. 2008)

Donnelly Const. Co. v. Oberg/Hunt/Gilleland, 677 P.2d 1292, 139 Ariz. 184 (1984)

Eichleay Corp., ASBCA No. 5183, 60-2 B.C.A. (CCH) ¶ 2688 (1960)

Estate of Stephen Puza v. Carbon County, 586 F. Supp. 2d 271 (M.D. Pa. 2007)

Federal Ins. Co. v. Winters, 354 S.W.3d 287 (Tenn. 2011)

Hewett-Kier Const., Inc. v. *Lemuel Ramos and Associates, Inc.*, 775 So. 2d 373 (Fla. Dist. Ct. App. 2000)

Indemnity Ins. Co. v. *American Aviation*, 891 So. 2d 532 (Fla. 2004)

Mesa Vista v. *Cal. Portland Cement Co.*, 12 Cal. Rptr. 3d 863 (Ct. App. 2004)

Moransais v. *Heathman*, 744 So. 2d 973 (Fla. 1999)

Northeast Clackamas County Electric Cooperative, Inc. v. *Continental Casualty Company*, 221 F.2d 329 (9th Cir. 1955)

Tirella v. *Am. Props. Team, Inc.*, 145 A.D.2d 724, 535 N.Y.S.2d 252 (App. Div. 1988)

United States v. *Spearin*, 248 U.S. 132, 39 S. Ct. 59, 63 L. Ed. 166 (1918)

2 General overview of BIM and VDC

2.1 General overview

In the preceding chapter, the reader was introduced to an overview of the well-established types of construction claims. This chapter now offers a broad introduction to BIM/VDC. The objective of this chapter is to outline the basic capabilities and functionality of BIM/VDC tools and workflows, as well as the general shape of BIM/VDC implementation and adoption by designers and contractors. The goal is to provide the armature of these new tools and processes upon which traditional claims based on legacy means and methods can be critically assessed.

As umbrella terms, Building Information Modeling and Virtual Design and Construction cover the software, hardware, and integrated and collaborative processes of risk mitigation in the design, construction, and operation of a building/facility project. BIM/VDC is not any one particular piece of proprietary software, any one particular type of digital file produced by software, or any one specific digitally enhanced process. Nor is BIM "computer aided design (CAD) 2.0" suggesting an incremental improvement over traditional CAD processes.

BIM/VDC is a fundamental shift in both the tools and the methods of design, construction coordination, and operation of buildings/facilities. Given its relevance to all stakeholders throughout the design–construct–operate continuum multiple definitions of BIM/VDC abound. However, the definition drafted by the National Building Information Modeling Standards (NBIMS) Committee is generally recognized as unifying.

A Building Information Model (BIM) is a digital representation of physical and functional characteristics of a facility. As such it serves as a shared knowledge resource for information about a facility forming a reliable basis for decisions during its life-cycle from inception onward. A basic premise of BIM is collaboration by different stakeholders at different phases of the life cycle of a facility to insert, extract, update or modify information in the BIM process to support and reflect the roles of that stakeholder. The BIM is a shared digital representation founded on open standards for interoperability.[1]

Over the course of the last decade BIM use and adoption in the United States has steadily transitioned from theory into practice and mainstream execution. Evidence corroborating this transition includes, for example, BIM-specific form contract documents by the American Institute of Architects® (AIA) and ConsensusDocs™. These form contracts, those of the AIA already in their second generation, provide evidence of the evolving standard of care with regard to BIM/VDC because they reflect market demands to explicitly address the business and risk aspects of these tools and processes. (These documents are reviewed at length in Chapter 3.)

Similarly, there are any number of industry coalition and publically available BIM guidelines and standards including those by Pennsylvania State University,[2] the States of Ohio and Wisconsin,[3] and the previously cited National Institute for Building Science *National BIM Standard – United States v2*[4] that provide the AEC industry with planning and execution resources. Likewise numerous books, refereed academic journal articles, and general market reports continue to document the paradigm shift in the AEC industry. For example, the seminal book, *BIM Handbook: A Guide to Building Information Modeling for Owners, Managers, Designers, Engineers, and Contractors* by Chuck Eastman, et al. is already in its second edition.[5] Multiple journal articles have investigated specific issues, including BIM/VDC practice assessments for mechanical, electrical, and plumbing (MEP) engineers and contractors,[6] and potential long-term uses for BIM in facilities management.[7] Additional research has explored frameworks for accurately measuring the benefits of BIM/VDC supported by performance metrics based on case studies.[8] Additionally, the National Institute of Building Sciences used to publish the *Journal of Building Information Modeling*, the content of which has now been amalgamated into the bi-monthly *Journal of the National Institute of Building Sciences*.[9]

From a market perspective, McGraw-Hill Construction continuously publishes AEC reports documenting the BIM/VDC transformation. Recent updates to their five-year compendium tracking BIM from 2007 through 2012 found a more than triple increase in the percentage of companies using BIM between 2007 and 2012. The same report found that BIM users experienced significant benefits in utilizing BIM including better profits, less re-work, reduced project duration, and fewer claims.[10] With BIM research beyond North America, McGraw-Hill also recently issued *The Business Value of BIM for Construction in Major Global Markets*.[11]

Sources such as these, along with innumerable others, support a thesis that the primary driver and value-add for the adoption and use of BIM/VDC is the ability to spatially coordinate digitally in three dimensional (3D) space during both design and pre-construction. Commonly known as "clash detection," this BIM/VDC process has been a long-time objective of the AEC industry and has now proven its potential to deliver cost savings to multiple stakeholders on a project.[12] Those potential cost savings are realized through the early detection and resolution of both design and constructability issues prior to the issuance of contract drawings and commencement of field-work, respectively.

Essential to the BIM/VDC process are multiple, individual, discrete Building Information Models (BIMs) or contributions from different stakeholders. As noted in the NBIMS definition above, BIMs are discipline-specific, data-rich, digital representations of the physical, functional characteristics of a building or facility. While most "cash value" discussions around BIM/VDC relate to the low-hanging fruit of 3D clash detection, it is critical to note that in addition to 3D geometric representations, BIMs may include two-dimensional (2D) drawings. These 2D drawings may be derived from, and integral with, the 3D geometric representations. In addition to 2D and 3D representations, a BIM might also contain pertinent associated data or specification information. For example, the light fixture is not only represented in 3D which supports above-ceiling coordination, but also contains lamp and ballast information supporting energy analysis, as well as direct links to a manufacture's website containing full specifications to support long-term facilities inventory needs.

For the longer arc of this research in terms of disputes and claims, a brief note here on the relationship between 3D BIMs and 2D contract drawings. This is a topic that receives significant attention in most BIM discussions. (This topic will be discussed throughout subsequent chapters, e.g. in Chapter 4 Section 4.4 Technical – 2D–3D conversion.) That attention is warranted for, as seen in the previous chapter, the alleged deficiencies in quality and completeness of contract documents is often a primary issue in claims. Parties trying to resolve such claims are often trying to answer questions such as, "Are the documents suitable for their intended purposes, that is, pricing, bidding, or construction?", "Are the documents reasonably complete, coordinated, and internally consistent?" or "Has adequate interdisciplinary design coordination been effectively implemented?"

While various opinions exist regarding the relationship between BIMs and 2D contract drawings, evidence supports a best practice wherein 3D BIM is the basis for the production of 2D printed contract documents. 2D drawings are extractions from the BIM, and the market leading BIM tools allow for bi-directional associativity between BIM and drawing. Basically, a change to the 3D model updates the 2D drawing and vice versa. This best practice has been noted in various industry sources and has been a contractual requirement for project delivery by certain owners since the release of their inaugural BIM guidelines.[13] From a legal perspective, the ability to designate a BIM as a contract document has been provided for in industry form contract documents like the ConsensusDocs™ 301 BIM *Addendum*, should the parties choose to do so. This is not to disregard the production techniques (some quite nuanced) involved in getting from BIMs to 2D contract drawings. Nor does this suggest that, at present, everything is modeled in 3D. 2D line work will certainly remain relevant for the foreseeable future. The note here simply recognizes the evidence supporting a current best practice and state-of-the-art where BIM is not something done in addition to, or separate from, the production of 2D printed, contract drawings.

In summary, BIM/VDC is a comprehensive process for risk mitigation utilizing 3D geometric, data-rich, digital representations of a building or facility. Multiple sources provide evidence of the ability of BIM/VDC to improve visual communication via 3D clash detection, thereby enabling faster, more comprehensive resolution of issues that might negatively impact design, coordination, or construction installation sequences. The benefits of clashing for both designers and contractors have, in turn, the potential to reduce requests for information (RFIs) and change orders since any clashes identified might otherwise go un-identified and un-resolved until it is too late and the affected parties incur potential additional cost and/or delay. Additionally, the primacy of 3D notwithstanding, 2D content still plays a role in a BIM/VDC enabled project. However, in best practice the production of 2D printed, contract drawings during design is not something done in addition to, or separate from, the use of BIM. Model and drawing are intimately connected as supported by examples of case studies, multiple BIM guidelines, and at least one flexible form contract that enables a BIM to be designated a contract document.

2.2 Preliminary definitions

The expansive and growing body of BIM knowledge and source materials has produced no shortage of new terms or acronyms. For the purpose of this chapter, which is to provide a general introduction and orientation to BIM/VDC tools, workflows, and processes, the following definitions will apply. These choices reflect an attempt to consolidate and recognize existing successful industry efforts at setting common BIM/VDC language without suggesting endorsement of any one of those particular efforts by using its specific terms as part of this overview. In later chapters which specifically review or cite source documents and their associated terms, those actual terms and meanings will be used.

> BIM(s) – At a minimum, 3D computer-generated models produced by relevant discipline stakeholders during both the design and construction phases of a project. BIMs may include entire discipline systems, for example a complete heating, ventilation, air conditioning (HVAC) system across the entire floor of a laboratory building, or discrete discipline elements, for example a single condensate drain line being installed as part of that larger system.

> Design BIMs (D-BIMs) – BIMs produced by the architect and their design consultants. Through the course of a project D-BIMs ultimately achieve a level of parity with traditional 2D construction documents.

> Compiled D-BIM (CompD-BIM) – The compiled yet distinct D-BIMs of the design team.

> Construction BIMs (C-BIMs) – BIMs produced by the construction team. As industry custom also variously designates them, C-BIMs might also be used interchangeably with "fabrication models" or "shop models" as these models achieve parity with traditional 2D shop drawings.

> Compiled C-BIM (CompC-BIM) – The compiled yet distinct C-BIMs of the construction team which typically also include D-BIMs.

2.3 Design BIM – general outline of tools and workflows

Design team professionals produce their respective D-BIMs using discipline-specific software tools. These software programs are sometimes referred to as "authoring" tools because users author, or create, digital BIM content (e.g. walls, doors, windows, ducts, pipe, valves, etc.) in the process of virtually building their D-BIMs. Distinct from BIM authoring tools are BIM/VDC project review and collaboration tools which are used for compiling, viewing, and "clashing" (a primary feature among others which will be discussed below) BIMs, but not creating BIMs or BIM content. For example, BIM project review tools have historically been used by contractors to compile, view, and analyze BIMs authored by others. Notwithstanding the distinction between authoring and collaboration BIM tools, many design teams are utilizing both BIM authoring and collaboration tools to deliver more accurate and complete contract documents. Furthermore, in addition to geometric clash tools, certain project review tools such as Solibri Model Checker enable "semantic" clashing that supports code-compliance analysis (e.g. fire egress).[14]

The marketplace has produced a number of BIM authoring tools, each with specific features, strengths, and weaknesses. The market leading BIM authoring tool is the Autodesk® Revit® platform.[15] Accordingly Revit is used as the vehicle for discussion in these orientation sections. Revit has three distinct, discipline-specific feature sets, namely: Revit Architecture, Revit Structure, and Revit MEP. D-BIMs produced in Revit Architecture, Structure and MEP are synonymous and can be linked together. (This process will be discussed in further detail below.) Furthermore, Revit allows intelligent linking of certain objects between and amongst D-BIMs, enabling, for example, one D-BIM to "know" when certain objects, such as a datum gridline location, has changed dimensionally in another linked D-BIM.[16]

The file extension produced by each version of Revit is an ".RVT". The analogy is to the pervasive ".DWG" file extension/type produced by the 2D CAD program Autodesk® AutoCAD®. Revit .RVT files can be used to extract and/or export multiple different files types, including: 3D.DWGs, 2D.DWGs and .PDFs for consumption by other software programs. This enables the extraction and production of 2D printed, contract drawings to be delivered in typical .DWG and/or .PDF file formats, where required.[17]

Revit Architecture provides core functionality, features, and BIM components for architects, for example walls, doors, windows, and so on. In Revit parlance these components are referred to as "families" and constitute the core building blocks of D-BIMs produced in Revit. Typically, families contain 3D and 2D geometric information about each component, as well as additional pertinent technical and/or performance data.[18] For example, a Revit door family might contain the 3D geometry to visualize that door correctly when seen from any 3D view or angle within the model. Similarly, when a 2D interior elevation needs to be drawn – in actuality, extracted directly from the BIM – Revit will automatically produce the required elevation view of the door to be placed on the 2D construction drawing. If required, application of any necessary additional detail or 2D line work will depend upon drawing scale and requirements, but a basic connection is made between the BIM object in space and its presentation on any associated 2D drawings.[19] Furthermore, if the door is deleted from the 3D model, the door is consequently and simultaneously deleted from any and all 2D drawings and/or schedules on which the door is drawn/shown. The same holds true if the door is deleted from an extracted 2D drawing or graphic schedule. Adding or removing an object to any view (2D, 3D, tabular) within the database accomplishes that task for all views of the database.[20]

In addition to geometric information, the door family might contain certain descriptive information or "parameters" such as fire-rating, frame-finish, head height, and so on. Doors with similar core characteristics will be grouped as named families – for example "single-flush" – and contain variants within that family, for example 30" × 80", 36" × 84", which are known as "types." Individuals' door types installed into a project become "instances" of those types, hence a project may have several dozen instances of a 30" × 80" single-flush door.[21] Revit "understands" door components to be just that – doors – and thus treats them with particular rules corresponding to their *door-ness*. For example, a door must be hosted by and reside within the confines of a wall. Similarly, when a Revit® D-BIM is queried for the number of instances of Single-Flush, 30" × 80" doors it contains, the software can quickly return a precise quantity as a result of the aforementioned family/type/instance hierarchy.[22]

The comprehensive information structure of component families, combined with the bi-directional associativity between D-BIMs and 2D printed drawings is what gives BIM its cohesive database structure in contrast to traditional, fragmented 2D CAD drawings. A typical 2D CAD program would describe, or represent, a door as a series of digital "lines on a page." That 2D CAD program, barring additional customization, would not for example, "know" that a door must sit within a wall, nor could it quickly schedule all instances of a door type. Nor would a 2D CAD program know what the door should look like when toggling between plan, section, or elevation views of the door for final placement on a 2D, printed contract drawing. Likewise, if the door is removed from a 2D plan view in which it appears, it would not automatically be removed from any and all other 2D drawings (e.g. elevations, sections) where that door is drawn/exists.

2.4 Design BIM – general outline of processes and procedures

In light of the general outline of BIM authoring tools above, a broad outline of the typical BIM/VDC processes and procedures typical of designers follows.

The architect and their design consultants will break up a project and produce multiple, individual D-BIMs to represent various aspects and systems of a project. For example, an architect might produce "core," "shell," and "fit-out" D-BIMs.[23] Depending on the scale and type of project, the architect might create additional D-BIMs corresponding to distinct levels, groups of levels, or programmatic zones/areas within the project, for example "retail," "ground-floor," or "tenant flrs.". This breaking up of a project into smaller, more manageable D-BIMs serves several purposes including streamlining design coordination sequences, address-ing the working/production needs of the architect so different designers don't need to actively work in the same BIM, and making file sizes smaller and more easily managed.[24]

If an architect of record is working collaboratively with a consulting architect in the production of construction documents they will establish rules for collaboration within and between the various D-BIMs each party will create both independently and collectively. Boundaries for respective design scope between the two design teams will be articulated. For example, the architect's BIM scope may not include ceilings, or exact placement of interior partitions in secondary spaces, but may be the responsibility of the consulting architect. If these duties are shared, additional organ-izing measures will be taken to define scopes of BIM work such as segregating ceilings along corridor/non-corridor distinctions, or some other method.

The relationship between architect and consulting architect requires estab-lishing ownership of "sheet views" in their respective D-BIM(s) as they relate to issued contract document set(s). Sheet views are/become the 2D, printed draw-ings of a contract document set derived from the BIM.[25] Thus, sheet views from the D-BIMs of both the architect and the consulting architect may be contract documents in the final issued sheet set. The architect and consulting architect may need to coordinate instances where they are co-authoring sheet views and each contributing substance to a single 2D, printed contract document. Planning between an architect and a consulting architect is critical not merely to properly mitigate any potential hardware issues when sharing models, but for basic project organization and administration in the production and issuance of contract docu-ments, and any subsequent bulletins, sketches, and so on under the responsible charge of the architect.[26]

As with the architect and consulting architect, the structural consultant might produce any number of corresponding structural D-BIMs. Similarly, the MEP design professional might produce corresponding, "mech," "plm-DR (drain)," "plm-SP (supply)," and "elec" D-BIMs. Lastly, additional design consultants such as, fire protection, and telecommunications might also produce multiple, appro-priately segregated D-BIMs.

Procedurally, the current state-of-the-art has architects and their consultants exchanging their respective BIM files on a highly regular basis, beginning as early as schematic design, and typically not less than once per week. The methods for exchange can vary widely depending on project teams, stretching from utilitar-ian file transfer protocols (FTPs), to more solution-specific server methods (e.g. Revit Server), and/or other types of project cloud collaboration and management software/systems.[27]

While the potential for other arrangements exists, it is typically the architect's responsibility to incorporate and maintain continuity between all of the numer-ous and various D-BIMs being produced by the collective design team. Current AIA digital practice documents designate the architect as the default project par-ticipant responsible for management of the models.[28] For the purpose of analysis within, the compiled D-BIMs of the design team are referred to as a Compiled

D-BIM (CompD-BIM). In Revit parlance, a compiled .RVT containing multiple other linked .RVTs might also be made a "Central File" which has certain performance characteristics as a master file.[29]

The major benefit of the CompD-BIM approach for the architect and their consultants is continuity and improved accuracy and quality in the production of their contract drawings. When all the various D-BIMs are brought together and linked, any one design professional can see the work and contributions of all other design team professionals, in coordinated 3D and 2D views (Figure 2.1). Historically, coordination of hand or 2D CAD drawings between and amongst design teams has been a fragmented and error-prone process because those design drawings were often produced in isolation and/or with patchy, static references to related drawings in a contract set (the so-called "x-ref'ing" of 2D background drawings). In contrast, a CompD-BIM presents a comprehensive and holistic view of all facets of a project.

In practice, BIM allows architects and their consultants to create any 2D plan, section, elevation, detail, and so on in essentially "real time" that automatically includes and displays all the elements of each linked D-BIM as currently modeled. For example, linking together the architect's "fit-out" BIM with the structural engineer's "struc" BIM, and the MEP designer's "mech" BIM would allow for the instantaneous creation of a 2D section drawing that accurately reflects ceiling height, bottom of steel, and ductwork – exactly as the architect and its consultants had designed and modeled/placed them in 3D space in each of their respective D-BIMs. Furthermore, should the architect subsequently adjust the ceiling height in his "fit-out" D-BIM, that change in ceiling height would be more clearly relayed to the structural and MEP engineers in the next exchange of D-BIM.

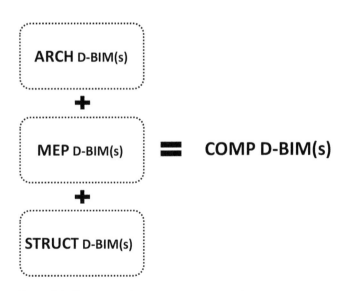

Figure 2.1 Design coordination process: general arrangement

While the above scenario does produce better continuity amongst the design team, there is the potential for a false positive, or design coordination to unrealistic dimensional thresholds or tolerances when it comes to real world construction. Accordingly, architects and their consultants typically use both authoring tools such as Revit and project review tools in concert.[30] As with BIM authoring tools, multiple project review tools and methods exist. A market leading project review solution has been Autodesk Navisworks® (a desktop solution).[31] Recently, Autodesk's BIM360 Glue® solution, which is a cloud-based Software as a Service (SaaS) delivery model appears to be the logical progression of the Navisworks solution.[32] Accordingly, these tools will serve as the vehicle for discussion here.

Neither Navisworks nor BIM360 Glue are tools for creating BIM content (i.e. doors, walls, etc.). Instead, these are tools for project review enabling compiling, viewing, and performing clash detection between and amongst various BIMs. An analogy within the industry might be to consider Navisworks and BIM360 Glue similar to PDF-*esque* tools and environments for quickly compiling, reviewing, and sharing information in a light-weight, easily shared document, or, in the case of BIM360 Glue, a cloud-based environment.[33]

A reason architects now utilize both Revit and Navisworks/BIM360 Glue is because while the former allows for the linking of multiple D-BIMs, thereby supporting the collaboration and scenarios described above, the latter tools are purpose-built for quickly discovering clashes between and amongst building systems that might not be visually obvious or otherwise easily identifiable and, which if left un-addressed, might lead to more serious constructability issues downstream. Collectively, this enables a designer to bring efficiency and accuracy to the production of their 2D contract documents by linking together the systems-specific D-BIMs of their consultants, thereby mitigating risk by providing greater accuracy and assurance that those systems as designed – and depicted on the 2D contract documents – will fit in approximately the locations shown.

For example, "Does the ductwork merely clip some kicker brace steel, or are there major areas of congestion with ductwork passing through beams as designed?", "Do plumbing drains pass through ductwork but show space for re-alignment during trade coordination, or do they pass through ducts in a structural bay already heavily congested?", or "Does a given mechanical pitched piping drain run clear of ceiling heights as designed?" With the combined use of BIM authoring tools and project review solutions an architect and his consultants can more adequately address questions such as these because the MEP engineering consultant begins the process by modeling/authoring in 3D, fully aware of space and fit constraints, as opposed to providing suggested layouts based on center-line approximations in 2D drawings.

With the current state of the BIM/VDC art, contract drawings produced by design teams utilizing BIM/VDC are understood by all members of a project team to deliver, at a minimum, a higher level of coordination within and across professional disciplines and building systems than is achievable with traditional, 2D CAD or paper-based methods for producing construction documents.[34]

The reasons for this are two-fold: the practical benefits derived from linking various D-BIMs and associated design-side clash detection processes as described above, and because with BIM/VDC the coordinated, 3D D-BIMs serve as the fundamental backdrop for the creation and production of 2D printed drawings. Once established, there is a permanent, two-way connection between the 3D D-BIMs and associated 2D drawings they produce. Recall the aforementioned discussion of bi-directional associativity. A change to a condition or element in the 3D D-BIM results in a corresponding change to the 2D drawing(s) describing that condition or element of the D-BIM, and vice versa.

2.5 Construction BIM – general outline of tools and materials

Similar to the process and workflow of design professionals creating multiple D-BIMs and linking them in a CompD-BIM for the resolution of design coordination issues prior to the issuance of 2D contract documents, there is a subsequent and similar process for construction teams. Subcontractors develop and compile various C-BIM/shop models to facilitate coordination, sequencing, scheduling and, where required provide as-built C-BIMs.

Current best practice calls for each subcontractor to develop a C-BIM of their scope of work in advance of the subcontractor's shop drawing submittal and subsequent on-site construction. At present, most contractual requirements for C-BIMs do not typically relieve contractors from their requirements to prepare traditional shop drawings. Subcontractors base their C-BIMs on the 2D contract drawings issued by the design team.[35] Many projects also make available some or all of the design team D-BIMs for use as reference by subcontractors. For example, the Associated General Contractors (AGC) acknowledges the benefit to construction teams of receiving constituent D-BIMS, or the entire CompD-BIM.[36] Likewise, industry form documents from the AIA and ConsensusDocs provide means of facilitating exchange and establishing rules of D-BIM and C-BIM development, reliance, and authorized use between design and construction teams. (These form documents are fully dissected in Chapter 3.) Where D-BIMs are provided to construction teams, contract language usually establishes that they are being provided as reference documents, with the 2D contract documents taking precedence.

On projects where a contractor also has a fixed-fee contract for pre-construction services, the contractor might typically conduct a pre-construction review of a design team's CompD-BIM at regular intervals prior to the issuance of bidding and contract documents as part of their normal estimating and constructability review processes. In such cases contractors might review the CompD-BIM to corroborate quantity counts derived from their typical paper-based, or on-screen take off methods for preparing estimates and/or share the corroborated quantities with bidding subcontractors. Likewise, they review and utilize the CompD-BIM in planning site safety, logistics, and major equipment (e.g. tower crane) staging, sequences, and movements. Contractors typically perform regular and periodic clash detection tests on the CompD-BIM prior to issuance of bidding and contract documents in order to validate the completeness of design integration of the design team. This allows the contractor to identify problem areas of systems congestion that may require additional attention during trade coordination and take appropriate preventative action. Such a process supports a contractor's implied "duty to inquire" with respect to any patent ambiguity in bidding drawings. To address larger concerns regarding this type of analysis by contractors, current form documents are clear that BIM/VDC is not intended to alter traditional notions of the architect as the person in responsible charge of the design, nor transform contractors into designers.[37]

As with the above discussion of BIM authoring tools for designers, there are any number of authoring tools for specialty trade subcontractors and fabricators to select from according to need and preference. For example, MEP subcontractors can create their 3D component models in trade-specific software such as Quickpen®, EastCoastCAD®, or SprinkCAD®. Cast-in-place concrete trades might use Tekla®, or Revit Structure®. Many of these trade-specific software programs are often "add-on" products for traditional CAD programs such as Autodesk® AutoCAD®. In many cases these trade-specific AutoCAD® add-on programs tie in directly and interface with a subcontractor's shop floor machinery for cutting, bending, and so on of materials during fabrication.[38] Thus, BIM/VDC workflows inherently support opportunities for pre-fabrication of constructed elements. Pre-fabrication, long a vision of some in the AEC industry, is proving a viable means of quality and cost control.[39]

The current state-of-the-art typically requires subcontractors to deliver their component models as 3D .DWG files, in addition to their contractually required 2D drawings. Within these 3D .DWG files, subcontractors typically model all the physical objects and space requirements necessary to coordinate their scope of work with other trades. In certain instances, subcontractors might use abstractions to represent objects in their scope of work, for example a box representation in place of a specific light fixture with distinct internal geometry.

Current best practices call for subcontractor C-BIMs to be broken down by trade, level, and zones, as necessary. The break-down of the component models along these lines is done to correspond with the general sequence of coordination as determined and set by the contractor. Similar to the design team workflows and processes, the breaking down of C-BIMs on the Project into discrete trades and levels brings efficiency to the process by streamlining coordination and sign-off, ensuring all component models are kept distinct and separate, and supporting manageable file sizes. The various C-BIMs/component models/shop models from each subcontractor are compiled into what has been termed for this section a CompC-BIM. The CompC-BIM is typically created in BIM project review software such as Autodesk Navisworks, or BIM360 Glue.

Thus, similar to the design-side clashing conducted by an architect, Navisworks and/or BIM360Glue allows the C-BIMs developed by each subcontractor to be superimposed over one another in 3D and remain distinct. Utilizing the built-in features and functionality of project review solutions the construction team can locate and isolate clashes and collisions between trade-specific shop models in 3D space. The team, typically led by the contractor can then use the management features of the solutions to track any clashes and collisions through to resolution prior to sign-off and the commencement of field-work. This management role is also often delegated by the contractor to the mechanical contractor given that trade's primacy in coordination.

In addition to the benefits of 3D clash detection there are the so-called "other D's" of a BIM/VDC enabled process. For example, 4D refers to the fourth dimension, time, and the potential for a BIM to be linked with certain construction scheduling software. In overly simple terms, the individual tasks of the construction schedule, with their planned and actual start and end dates, can be applied to elements within the BIM. As a basic example, imagine a schedule that includes tasks for: installing curtain wall panel "A" starting on Monday, finishing on Tuesday; curtain wall panel "B" start Tuesday, finish Wednesday; curtain wall panel "C" start Wednesday, finish Thursday, and so on. The construction scheduling software is then linked to the applicable BIMs such that the corresponding start and end dates from the schedule are applied to the 3D curtain wall panels A, B, and C. Having established such a connection, the BIM can be animated to play back a time-lapse film of the installation occurring over the course of the week. Likewise, for planning purposes a contractor could select any date within the schedule – past or future – and have a corresponding visual snapshot of the BIM. Features such as these allow the contractor to query the model and visualize sequences concerning, for example, site logistics and systems installation sequences. (As will be discussed in Chapter 3, while not BIM/VDC specific, there is case law precedent regarding construction scheduling software.)

There is also "5D" or cost-related information that can be supported by BIM. While at present no BIM authoring tool is expected to also be a complete construction estimating tool, there are any number of workflows and/or stand-alone estimating and cost analysis tools that can leverage BIM content in pursuit of cost control.[40] As previously described, the database nature of BIM platforms allow for quick queries of quantity information – for example, "How many 2hr fire-rated doors are on the 2nd floor of this project?" Furthermore, model content may be organized, as is the case with models authored in Revit, according to the Construction Specification Institute (CSI) UniFormat® system. Revit also organizes keynote annotation under CSI's MasterFormat® system (1995 and 2004).[41] Likewise, UniFormat® is also used as the organizing methodology for model content creation and ownership between project participants using the AIA E202-2008. While there are distinctions and considerations regarding the use of any classification systems, the point is that objects within the model are organized in such a manner as to facilitate the preparation of estimates. Again, this is not to suggest that preparing a detailed construction estimate is a "one-click solution" from a BIM. Construction estimates are as much art as science. But, the framework of BIM content as a database and the ability to extract quantities, assign specific parameters to objects, and/or establish relationships between BIM content and outside cost information enhances traditional workflows.

In addition to schedule and cost components, contractors are also increasingly utilizing mobile and cloud-based technologies to manage construction activities in the field. The mental picture of the construction superintendent walking the job site with a roll of drawings under his arm and a pencil in his ear is being forced to share the frame with an iPad and mobile phone. For example, technology such as Autodesk BIM360Field® (the complementary other half of BIM360Glue as described above) brings anytime, anywhere access of project data, including BIM content, to the field. Use cases might include a subcontractor utilizing the camera on an iPad during installation of equipment to scan a barcode on that equipment and immediately be presented with a checklist of tasks to complete, data entry fields, equipment specifications, and a link to the BIM model of that exact piece of equipment. Field layout technologies also enable x,y,z coordinates from a BIM to inform the physical installation locations of construction work (e.g. hangers, sleeves in floor slabs) via global positioning systems (GPS). Conversely, scan and survey data points from the field can be imported back into models to verify accuracy. These tools for so-called "field BIM" are meant to further leverage the clashing process used to identify spatial coordination issues during the design or pre-construction phases of a project. For the same reason that companies like Federal Express® or UPS® manage the millions of packages they ship everyday by giving field personnel mobile device scanners that provide real-time status and feedback, construction projects are seeing the benefit of managing the millions of moving parts in complex construction projects through the use of mobile and cloud technologies.[42] Recent project-based research suggests that field-based technologies such as these can yield significant returns on time and productivity.[43]

2.6 Construction BIM – general outline of processes and procedures

The contractor will set a schedule for BIM coordination meetings and compiling the most recent subcontractor C-BIMs into the CompC-BIM. The contractor will create a "clash report" which identifies clashes between trades in the CompC-BIM (Figure 2.2). Clashes between trades typically adhere to the following hierarchy and sequencing:

1   structural steel and architectural elements
2   mechanical ductwork
3   gravity plumbing and mechanical systems
4   pressure plumbing and mechanical systems
5   electrical
6   fire protection.

(A brief note here regarding the contractor's BIM coordination process. On most projects there comes a point during coordination when the structural D-BIM of the structural design engineer will become insufficient in detail for full trade coordination and will give way to the more detailed C-BIM of the steel fabricator.)

Clash reports are distributed to the relevant members of the construction team prior to the next BIM coordination meeting. The clashes within and between

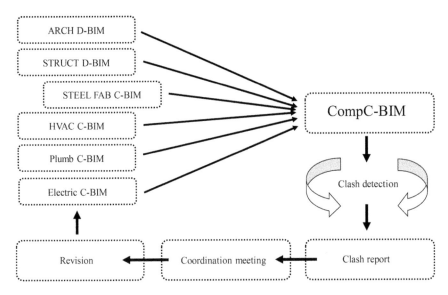

Figure 2.2 Trade coordination process: general arrangement

Source: Diagram after *AGC Contractors' Guide to BIM Edition 1* (2006).

systems that are identified in the clash report become the focus and agenda of the next BIM coordination meeting. BIM coordination meetings might be held in a "BIM room." BIM rooms of high quality typically contain multiple (minimum two) large screens for display of CompC-BIMs, shop models, D-BIMs, 2D contract drawings, and so on.

The increasing availability of cloud-based project review tools has added additional efficiencies to the traditional "compile-clash-report" workflow described above. With a cloud-enabled solution, such as the aforementioned Autodesk BIM360Glue, the contributions of various stakeholders to the CompC-BIM can occur in essentially real time, without the bottle-neck of requiring the contractor to manually gather the constituent BIMs and then process clashes in the CompC-BIM on a weekly (or more) basis. Cloud-based solutions such as BIM360 Glue also further enhance the compile-clash-report workflow by offering bi-directional association between the CompC-BIM and the constituent BIM in the native authoring tool(s) (i.e. Revit®). As noted by Eastman and others, the early lack of project review tools to offer bi-directional association created a static process in an otherwise efficient and fluid workflow.[44] With new bi-directional features, project participants can more easily streamline the "round-trip" process of compiling models, running clashes, and preparing and distributing clash reports.

This BIM-based coordination process of: (1) federating (Navisworks)/cloud gathering (BIM360Glue) component models, (2) running/automating a clash report, (3) issuing/auto-notifying the clash report to appropriate stakeholders, (4) holding a BIM coordination meeting, and (5) making necessary adjustments to C-BIMs is an iterative process dependent upon project complexity. Best practice typically requires between four and six BIM coordination meetings before an area is signed off. The introduction of bi-directional associations might change these dynamics and shorten cycles even further.

Once an area is signed off, final C-BIMs as they exist in the CompC-BIM are understood as contracts for space. By modeling all the necessary components of their scope and then submitting those components to the rigor of the clash process, subcontractors reserve their space for installation. Accordingly, when a conflict arises in the field post-sign-off, the subcontractor who did not reserve space for the components in their scope's C-BIM, must yield to those subcontractors who did reserve space by completely modeling all their necessary components. Components not modeled will be installed after those that are. Likewise, any re-work costs associated with incomplete and untimely C-BIM are borne by the author of those incomplete and untimely models.

2.7 Summary

Numerous sources reveal significant BIM/VDC adoption and usage in the United States. A continuously growing body of evidence demonstrating real benefits drawn from practical use cases on actual projects has begun to deliver reliable performance metrics. For architects, the existing authoring and project review tools, workflows, and processes enable more efficient design decision making and coordination. The bi-directional associativity between models and drawings directly supports delivery of better coordinated contract drawings. Likewise, contractors utilizing BIM/VDC tools and processes are able to better mitigate risk in project delivery. The benefits of visualization and clash detection, amongst others, have been shown to directly impact a contractor's ability to manage acceleration, avoid delay, and mitigate disruptions. For example, in delivering pre-construction services, a contractor's ability to interrogate an architect's CompD-BIM, even with full disclaimer and restriction, cannot help but enhance their implied duty to inquire.

Given such advances and achievements in an otherwise fragmented and, as some would call it, "broken" industry, how could claims and disputes continue to flourish? Perhaps they will. Perhaps they won't. Regardless of outcome, the thesis here is that the industry can only benefit from an analysis of these advancements against past issues. Considering if these new solutions might find themselves bringing past issues along with them, or creating new ones entirely, is a goal of the next chapter.

Notes

1 National BIM Standard – United States v2 (2012) 149. Accessed on February 22, 2014. http://www.wbdg.org/bim/nibs_bim.php#nbims.
2 Computer Integrated Construction Research Program. "BIM Project Execution Planning Guide – Version 2.1." (May, 2011). The Pennsylvania State University, University Park, PA, USA.

3 *Supra* 3.1.3.3, 3.1.3.4 .
4 *National BIM Standard – United States v2.*
5 Eastman, C., Teicholz, P., Sacks, R., & Liston, K. *BIM Handbook: A Guide to Building Information Modeling for Owners, Managers, Designers, Engineers and Contractors.* 2nd ed. Wiley (2011).
6 Boktor, J., Hanna, A., & Menassa, C. C. "State of Practice of Building Information Modeling (BIM) in the Mechanical Construction Industry." *Journal of Management in Engineering* 30.1 (2013): 78–85.
7 Becerik-Gerber, Burcin, et al. "Application Areas and Data Requirements for BIM-Enabled Facilities Management." *Journal of Construction Engineering and Management* 138.3 (2011): 431–442.
8 Barlish, K., & Sullivan, K. "How to Measure the Benefits of BIM—A Case Study Approach." Unpublished thesis. Arizona State University (2011). Also in *Automation in Construction* 24 (2012): 149–159. (Un-reviewed by author here.)
9 *Journal of the National Institute of Building Sciences.* Accessed on November 30, 2014. http://www.wbdg.org/references/jnibs.php.
10 McGraw-Hill. *The Business Value of BIM in North America: Multi-Year Trend Analysis and User Ratings (2007–2012).* Bedford, MA: McGraw-Hill Construction (2012).
11 McGraw-Hill. *The Business Value of BIM for Construction in Major Global Markets.* Bedford, MA: McGraw-Hill Construction (2014).
12 See, for example, for early research: Tatum, B., & Korman, T. *MEP Coordination in Building and Industrial Projects.* CIFE Work (1999).
13 See, for example, Eastman et al. *BIM Handbook,* Indiana University BIM Guidelines, NYC Department of Design and Construction BIM Guidelines.
14 Solibri. Accessed on December 20, 2014. http://www.solibri.com/products/solibri-model-checker/.
15 Eastman et al. 77.
16 http://help.autodesk.com/view/RVT/2015/ENU/. Accessed on November 30, 2014.
17 Ibid.
18 Ibid.
19 Ibid.
20 Ibid. See also Eastman et al. 78.
21 Ibid.
22 Ibid.
23 Ibid. See also, *Model Performance Technical Note.* Accessed on November 30, 2014. http://download.autodesk.com/us/revit/revit_performance/autodesk_revit_2014_model_performance_technical_note.pdf.
24 Ibid.
25 Ibid.
26 Ibid.
27 Ibid.
28 AIA E203-2013 *Building Information Modeling and Digital Data Exhibit.* §4.8.1.
29 http://help.autodesk.com/view/RVT/2015/ENU/
30 Eastman et al. 239.
31 Ibid.
32 http://www.autodesk.com/products/bim-360-glue/overview.
33 http://help.autodesk.com/view/BIM360/ENU/.
34 For example, Eastman et al. 21. McGraw-Hill. *The Business Value of BIM in North America: Multi-Year Trend Analysis and User Ratings (2007–2012).*
35 See Associated General Contractors. *AGC Contractors' Guide to BIM* (2005).
36 Ibid.
37 *Supra* Chapter 4.1.
38 See, for example, Eastman et al. Chapter 7.

39 See Kieran, S., & Timberlake, J. *Refabricating Architecture*. McGraw-Hill (2003).
40 Eastman et al. 277.
41 http://help.autodesk.com/view/RVT/2015/ENU/.
42 See Kieran and Timberlake.
43 Moran, Michael Stephen. "Assessing the Benefits of a Field Data Management Tool." Unpublished thesis. University of Delft (2012).
44 Eastman et al. 275.

References

American Institute of Architects. E202 (2008) *Building Information Modeling Protocol Exhibit*.

American Institute of Architects. E203 (2013) *Building Information Modeling and Digital Data Exhibit*.

Associated General Contractors. *AGC Contractors' Guide to BIM* (2005).

Autodesk BIM360 Field and Glue. Accessed on November 30, 2014. http://www.autodesk.com/products/bim-360-glue/overview.

Autodesk Model Performance Technical Note. Accessed on November 30, 2014. http://download.autodesk.com/us/revit/revit_performance/autodesk_revit_2014_model_performance_technical_note.pdf.

Autodesk Revit Help File. Accessed on November 30, 2014. http://help.autodesk.com/view/RVT/2015/ENU/.

Barlish, K., & Sullivan, K. "How to Measure the Benefits of BIM—A Case Study Approach." Unpublished thesis. Arizona State University (2011). Also in, *Automation in Construction* 24 (2012): 149–159. (Un-reviewed by author here.)

Becerik-Gerber, Burcin, et al. "Application Areas and Data Requirements for BIM-Enabled Facilities Management." *Journal of Construction Engineering and Management* 138.3 (2011): 431–442.

Boktor, J., Hanna, A., & Menassa, C. C. "The State of Practice of Building Information Modeling (BIM) in the Mechanical Construction Industry." *Journal of Management in Engineering* 30.1 (2013): 78–85.

Computer Integrated Construction Research Program. "BIM Project Execution Planning Guide – Version 2.1." (May, 2011). The Pennsylvania State University, University Park, PA, USA.

ConsensusDocs. 301 *Building Information Modeling (BIM) Addendum* (2008).

Eastman, C., Teicholz, P., Sacks, R., & Liston, K. *BIM Handbook: A Guide to Building Information Modeling for Owners, Managers, Designers, Engineers and Contractors*. 2nd ed. Wiley (2011).

Indiana University. *BIM Guidelines and Standards*. Accessed on February 22, 2014. http://www.iu.edu/~vpcpf/consultant-contractor/standards/bim-standards.shtml.

Journal of the National Institute of Building Sciences. Accessed on November 30, 2014. http://www.wbdg.org/references/jnibs.php.

Kieran, S., & Timberlake, J. *Refabricating Architecture*. McGraw-Hill (2003).

McGraw-Hill. *The Business Value of BIM in North America: Multi-Year Trend Analysis and User Ratings (2007–2012)*. Bedford, MA: McGraw-Hill Construction (2012).

McGraw-Hill. *The Business Value of BIM for Construction in Major Global Markets*. Bedford, MA: McGraw-Hill Construction (2014).

Moran, Michael Stephen. "Assessing the Benefits of a Field Data Management Tool." Unpublished thesis. University of Delft (2012).

National BIM Standard – United Statesv2 (2012). Accessed on February 22, 2014. http://www.wbdg.org/bim/nibs_bim.php#nbims.

NYC Department of Design and Construction BIM Guidelines. Accessed on February 22, 2014. http://www.nyc.gov/html/ddc/html/pubs/publications.

Solibri. Accessed on December 20, 2014. http://www.solibri.com/products/solibri-model-checker/.

Tatum, B., & Korman, T. *MEP Coordination in Building and Industrial Projects.* CIFE Work (1999).

Part II

Analysis

BIM/VDC form documents and guidelines, and legal concepts

3 Standard of care and workmanlike performance

Part I prepared the foundation for exploration here. Chapter 1 presented a general categorization of typical construction claims with an overview of some of the types of issues and damages sought in each. Chapter 2 presented an outline of BIM/VDC tools, workflows, and processes for both designers and contractors. Together these provided a view on to the current landscape of construction disputes, and a lens of new BIM/VDC tools and processes which might alter the shape of that landscape. The intent of this chapter is to present an analysis, via a broad sampling of primary and secondary sources, which might reasonably define minimum expectations for BIM/VDC standard of care and workmanlike performance as would likely be supported across the broadest geographic spectrum.

The structure of this analysis is in two parts. First, both concepts are viewed through examples of specific definitions and relevant clauses from several non-BIM/VDC form, or "boiler-plate," contract documents as prepared by the AIA and ConsensusDocs, a coalition commonly attributed with producing contractor-centric form documents. A small sample of court decisions regarding standard of care and workmanlike performance are also presented.

Second, atop this general footing and foundation sits a layer of BIM/VDC-specific analysis. This analysis will draw selected source documents from across three "bands" of industry knowledge spanning the past two decades.

1 General industry research – will provide a brief introduction and orientation to a small sample of select pieces of key, oft-cited research regarding BIM/VDC.
2 Guidelines and standards – presents an analysis of publically available BIM guidelines and standards that have emerged with BIM/VDC implementation and adoption.
3 Form contracts – will review recent and current BIM/VDC-specific form contract documents from both the AIA and ConsensusDocs.

Collectively, the source documents across three knowledge bands draw a representative cross-section of the emerging shape of the BIM and VDC maturity curve in the US and, hence, assist in defining baseline standards of care and workmanlike performance.

3.1 General concepts

In attempting to resolve a dispute where the standard of care of the design professional or workmanlike performance of the contractor has been alleged deficient, the parties involved will, generally speaking, take a look around to see what others similarly situated have done/are doing. In doing so, the parties are trying to determine what the standard of performance and conduct of a design professional or contractor facing the same or similar circumstances is, and whether or not the conduct at hand in their issue falls below such standard.

The standard of care against which design professionals are evaluated is, broadly speaking, different from that against which contractors are evaluated. Design professionals are required to hold a professional license (obtained through a lengthy education, apprenticeship, and examination process) when delivering their services. Contractors implicitly perform their services in a workmanlike manner and may, depending on jurisdiction, also be required to possess a business license or registration. This is not to suggest that contractors are not "professionals" in the broad sense of the word. A contractor may also be required to obtain a license only earned through a combination of journeyman experience hours and written examination(s) similar to the process for design professionals. However, a contractor is not legally burdened with making determinations concerning building code compliance in the design drawings, but instead executing the work in a workmanlike manner such that the completed work strictly complies with the contract documents as prepared by the design professional.

As previously noted, the clear separation between design and construction is a fundamental tenet of US construction law.[1] Beyond the US perspective, the general evolution of the legal separation between these two sides of the same coin can be engaged from any number of engrossing historical perspectives.[2] But for a truly "modern" perspective, no better dry toast is offered than a brief mention of the way each party insures themselves against losses. While the insurance marketplace offers a variety of instruments designed for distinct purposes, generally speaking architects must possess professional liability insurance, protecting them against "errors and omissions" in their professional practice. An error is customarily defined as a mistake in the design where the design element was either constructed or under construction and required retrofitting and/or placement of some component to correct the error. An omission is customarily defined as scope that was either missed and/or omitted by the designer in their construction documents (CDs) but was later discovered and added to the scope of work via a change order. Such policies normally exclude coverage for faulty construction work associated with a given project.

Conversely, contractors typically obtain, inter alia, builder's risk insurance which is designed to cover property during the course of construction, and commercial general liability (CGL) which covers third-party claims for bodily injury and physical property damage.[3] Generally speaking, architects insure their professional practice whereas contractors insure the given project and third-party claims for injuries or damage that may result from the construction of a project. The fundamental differences between the insurance instruments designed for architects and contractors, including the underwriting methodologies behind each, are a fertile opportunity for philosophical discourse on BIM and VDC within traditional design–bid–build contract arrangements, but any such thoughts are set aside here to continue with a more direct look at standard of care and workmanlike performance.

A survey of relevant excerpts from standard form documents issued by the AIA and ConsensusDocs will illustrate not only the professional/workmanlike distinction discussed above, but will also highlight the manner in which each respective organization defines those standards of measurement. First, a look at the definitions for the standard of care for the architect and design professional as the respective actors for each authoring body are named and defined.

In stating the standard of care to which an architect must perform, the AIA's self-designated flagship owner–architect agreement B101 (2007) reads:

> The Architect shall perform its services consistent with the professional skill and care ordinarily provided by architects practicing in the same or similar locality under the same or similar circumstances. The Architect shall perform its services as expeditiously as is consistent with such professional skill and care and the orderly progress of the Project.[4]

The B101 (2007) replaced two earlier AIA owner–architect documents, namely B141 and B151. While those documents imply a standard of care, the B101 (2007) is the first AIA owner–architect agreement to explicitly include a statement of what the standard of care is.[5] Also, 2007 was the first time owners participated in the revision process and may have been a contributing factor to the inclusion of the definition.[6] While the B101 (2007) was the first to include explicit standard of care language, that language closely approximates previous interpretation by the courts.[7]

In similar, yet distinct language the ConsensusDocs 240 *Standard Form of Agreement Between Owner and Architect/Engineer* (rev. 2014) states:

> The Design Professional accepts a relationship of trust and confidence with the Owner for this Agreement and will cooperate and exercise the skill and judgment required above in furthering the interests of the Owner. The Design Professional represents that it possesses the skill, expertise, and licensing to perform the Services. The Owner and Design Professional agree to work together on the basis of mutual trust, good faith, and fair dealing, and shall take actions reasonably necessary to enable each other to perform this Agreement in a timely, efficient, and economical manner. The Owner and Design Professional shall endeavor to promote harmony and cooperation among all Project participants.[8]

Next, a survey of example clauses and language concerning workmanlike performance of contractors. In the interest of variety, several different types of owner–contractor agreements are presented. Furthermore, each example agreement's accompanying terms and conditions (as separate or combined document(s) and as coordinated by the authoring body) are also examined.

The AIA C132 (2009) *Standard Form of Agreement Between Owner and Construction Manager as Advisor* states:

> The Construction Manager shall perform its services consistent with the skill and care ordinarily provided by construction managers practicing in the same or similar locality under the same or similar circumstances. The Construction Manager shall perform its services as expeditiously as is consistent with such skill and care and the orderly progress of the Project.[9]

The AIA's flagship recommended terms and conditions, A201 (2007) *General Conditions of the Contract for Construction*, offer additional insight into the practical realities of licensing and professional/workmanlike distinctions. For example, Section 3.1.1 states: "The Contractor is the person or entity identified as such in the Agreement and is referred to throughout the Contract Documents as if a singular number. The Contractor shall be lawfully licensed, if required in the jurisdiction where the Project is located." Further, Section 3.2 Review of Contract Documents and Field Conditions of the A201 (2007) contains clauses stating, in relevant parts: "it is recognized that the Contractor's review is made in the Contractor's capacity as a contractor and not as a licensed design professional, unless otherwise specifically provided in the Contract Documents."[10] And, "The Contractor is not required to ascertain that the Contract Documents are in accordance with applicable laws, statutes, ordinances, codes, rule and regulations, or lawful orders of public authorities, but the Contractor shall promptly report to the Architect any nonconformity discovered."[11]

Having taken a brief look at architect-centric documentation relevant to standard of care and workmanlike performance, we shift now to the perspective of another industry coalition, ConsensusDocs. Article 3 of the ConsensusDocs 200 *Standard Agreement and General Conditions Between Owner and Constructor - Lump Sum* (rev. 2014) in defining workmanship reads:

> The Work shall be executed in accordance with the Contract Documents in a workmanlike manner. All materials used in the Work shall be furnished in sufficient quantities to facilitate the proper and expeditious execution of the Work and shall be new except such materials as may be expressly provided in the Contract Documents to be otherwise.[12]

In addressing performance duties, ConsensusDocs 200 states, in part:

> It is recognized, however, that the Contractor is not acting in the capacity of a licensed design professional, and that the Contractor's examination is to facilitate construction and does not create an affirmative responsibility to detect errors, omissions or inconsistencies or to ascertain compliance with applicable laws, building codes or regulations.[13]

Finally, a very brief sampling of how some courts have addressed issues related to the professional standard of care for architects and workmanlike performance for contractors. As has been inferred throughout, the law of design and construction has a very long historical record. For example, in 2001 the Vermont Supreme Court cited a case from 1896 as the "leading case describing the negligence standard":

> The undertaking of an architect implies that he possesses skill and ability, including taste, sufficient to enable him to perform the required services at least ordinarily and reasonably well; and that he will exercise and apply, in the given case, his skill and ability, his judgment and taste, reasonably and without neglect. But the undertaking does not imply or warrant a satisfactory result.[14]

Echoing language utilized in the form contract excerpts above, a ruling from Washington State concerning the standard of care for architects affirmed that a trial jury was properly instructed that "an engineer or designer is guilty of negligence if he fails to apply the skill and learning which is required of similarly situated engineers or designers in his community."[15] Finally, an excerpt from *Melody Home Mfg. Co.* v. *Barnes* concerning the implied contractual warranty of good and workmanlike performance of a contractor: "We define good and workmanlike manner as that quality of work performed by one who has the knowledge, training, or experience necessary for the successful practice of a trade or occupation and performed in a manner generally considered proficient by those capable of judging such work."[16]

Having contemplated the "what" of standard of care and workmanlike performance, logic requires a consideration of the "how." Establishing if a design professional or a contractor has breached their respective duties typically requires expert testimony and that expert testimony is usually given by architects or engineers. Expert testimony can be given by other qualified people if they possess the requisite specialized knowledge, skill, training, experience, or education to address the issue.[17] For example, the Tennessee Court of Appeals has stated, "We have concluded that a person need not possess either a degree in architecture or an architect's certificate to be qualified to testify as an expert regarding the standard of care of architects and whether particular conduct violates that standard."[18]

With a footing in the general concepts of standard of care and workmanlike performance established, we return to the proposed intent from the beginning of this chapter: "How best to describe the professional standard of care for design professionals and workmanlike performance for contractors in the age of BIM and VDC?" What body of knowledge, what primary and secondary sources, what evidence – along with the first-hand experience of any retained expert – stands to collectively corroborate that experience and thereby bring to focus more clearly the current shape of the BIM/VDC maturity curve? The investigation into these questions is structured around a review of primary and secondary source documents from across the three "bands" of industry BIM/VDC knowledge, general industry research, guidelines and standards, and form contracts (Figure 3.1).

3.2 BIM/VDC source documents – general industry research

The source documents selected below present a sample of select research regarding BIM/VDC that have captured current states of practice and/or exhibited a level of influence on the AEC industry over the (roughly) past two decades. Individually and collectively, these documents provide a variable in reasonably describing the general progression of BIM/VDC capabilities and the advancing baselines for standard of care and workmanlike performance. The source documents/collections reviewed include:

- Center for Integrated Facility Engineering at Stanford University (1990s–present)
- US Department of Commerce *Cost Analysis of Inadequate Interoperability in the US Capital Facilities Industry* (2004)
- Associated General Contractors of America *Contractors' Guide to BIM – Edition 1* (2006)

3.2.1 Center for Integrated Facility Engineering at Stanford University

The stated mission of the Center for Integrated Facility Engineering (CIFE) is to be the world leader in academic research regarding VDC for the AEC industry. In pursuing their mission CIFE utilizes "translational research" which integrates on-site baseline observations of problems facing AEC practice, followed by a period of theoretical development of ideas to address those problems, before culminating in strictly designed validation studies to assess the utility of any proposed method or solutions.[19] A review of CIFE's publications database reveals a long tradition with many prescient observations and theories considering CAD/BIM/VDC development, use, and adoption.[20]

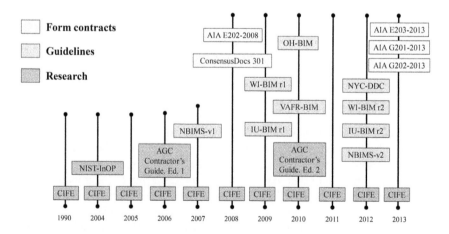

Figure 3.1 BIM/VDC standard of care: analysis across knowledge bands

For example, CIFE working paper #2, from January 1998, included amongst its predictions that CAD systems would soon automate basic design tasks, design databases would become the basis for cost estimates, 3D models would become standard practice for walkthrough analysis, and that as-built 3D deliverables would serve as the starting point for on-going facility management.[21] Similarly, other early research identified numerous benefits to technology in construction including utilizing computer models for coordinating equipment installation and maintenance access, as well as enhancing claims quantification.[22] By the mid-1990s CIFE was producing research that demonstrated the efficiencies and benefits to connecting 3D models with construction scheduling procedures.[23] By the start of the 2000s additional research was documenting the inefficient and error-prone process of traditional methods for coordinating MEP systems and pro-posing revised workflows that utilized 3D models and specialized clash detection software.[24] These examples represent a fraction of CIFE's collective publications.

In an effort to align their academic research with AEC industry business objec-tives, CIFE continues to propose VDC performance objectives for their member institutions. Current (2015) objectives include goals such as using VDC tools and methodologies to achieve zero hours lost to safety, and ≥ 95 percent of construc-tion budgeted items to be within 2 percent of cost.[25] The cumulative breadth and depth of CIFE's translational research rooted in validation testing, together with the current aggressive performance objectives for member institutions, provide a reasonable variable in any analysis that seeks to define the current standard of care and workmanlike performance regarding BIM/VDC.

3.2.2 US Department of Commerce Cost Analysis of Inadequate Interoperability in the US Capital Facilities Industry

In 2004 the US Department of Commerce via the National Institute of Standards and Technology (NIST) published *Cost Analysis of Inadequate Interoperability in the US Capital Facilities Industry* (NIST-InOp).[26] While not concerned with BIM/VDC in name, the research quantified the losses resulting from the siloed, patchy, and paper-based methods of traditional information exchange, management, and access in the design, engineering, construction, and facilities operations contin-uum. Interoperability costs were defined, essentially, by comparing the current fragmented and paper-based business activities (i.e. CAD) of the design-to-operations cycle against a hypothetical scenario wherein information exchange, management, and access was fluid (i.e. BIM). Significantly, the report's coun-ter-hypothetical scenario envisioned a design-to-operations process wherein design, engineering, construction, and operations information was entered into electronic systems once, and only once. Without marginalizing a complex prob-lem, one of the questions the NIST report sought to answer was, "How much is manual data re-entry costing the industry?"

The NIST report quantified an estimated $15.8 billion annually in total lost costs directly attributed to the fragmented nature of information exchange. $15.8 billion represented between 1 percent and 2 percent of revenue of the estimated

$347 billion of capital facilities work in place for 2002 (the year of the study).[27] At a specific, tangible level, $15.8 billion included lost costs such as the annual administrative re-entry costs that stakeholders might incur in resolving a request for information (RFI). For example, the subcontractor uses system or process #1 to generate an RFI, the contractor uses system or process #2 to manage that RFI, the architect then uses their own system or process #3 to respond/archive, with the owner using system or process #4 to log and track the RFI for their internal budgeting purposes. The cumulative effect of the four different systems in this example typically carries a cost associated with transition and/or duplication efforts between each system and process.

Disaggregated, annual stakeholder shares for the $15.8 billion were allocated as follows: architects and engineers – $1.2 billion, general contractors – $1.8 billion, specialty contractors – $2.2 billion and, owners – $10.6 billion.[28] Parsed at the square foot (sqft) level, the data revealed the existing fragmented processes to result in lost costs in excess of approximately $6/sqft during design and construction and $0.23/sqft/annum during facility operations.[29] Furthermore, the research self-diagnosed additional lost opportunity costs that the study was not designed to capture. Thus, the given $15.8 billion annual loss estimate was likely short of a larger true mark.

In quantifying (even if undervaluing) the cost and waste realities of "business as usual" for the capital facilities industry in 2002 the NIST-InOp validated the "gut feelings" of many designers, contractors, and owners. The report likely ranks as the most oft-cited authority in applicable literature.[30] As a source document tracing the arc of BIM/VDC standards it provides a valuable metric against which subsequent academic research and commercial project usage of BIM/VDC can be compared.

3.2.3 *Associated General Contractors* Contractors' Guide to BIM – Edition 1

Published by the Associated General Contractors of America in 2006, *The Contractors' Guide to BIM Edition-1*[31] (AGC-BIM) had the stated purpose of assisting contractors "who recognize this future [BIM/VDC] is coming and are looking for a way to start preparing themselves."[32] Further articulating their position the AGC-BIM dispelled any suggestion that BIM/VDC was only relevant to large projects, and/or only for use by large contractors. The AGC-BIM stated as fact that, "The benefits of using BIM on all projects, regardless of size . . . are being proven by contractors today" and that those benefits are likewise "being seen by contractors of all sizes."[33] With their over-arching position established, the AGC-BIM addressed a number of critical BIM issues for contractors to consider including: the fundamentals of the BIM process, software and technology, clarification of responsibilities, and risk management.

AGC-BIM defines BIM as both the data rich models representing a facility (noun) and the use of those models to simulate construction and operation of the facility (verb).[34] Regarding fundamental process, the AGC-BIM then orients readers to the notion of compiling design models prepared by designers, together with shop drawing/models from subcontractors as a "composite model" which retains the fidelity of each original contribution.[35] The AGC-BIM also provides performance criteria and budget considerations (in 2005 dollars) with respect to BIM software tool selection.[36]

An issue addressed repeatedly throughout AGC-BIM is the distinction between "2D Conversions" versus "3D Designs." As logically inferred, 2D-to-3D conversion describes the scenario in which 2D design documents are provided by the design professional to the contractor who then must convert those 2D documents into 3D in order to begin extracting a benefit (primarily systems coordination) from the BIM. At the time of publication, the 2D-to-3D conversion issue was in early contemplation as the AEC industry collectively adjusted to BIM. The AGC-BIM suggested that it would be five to ten years before "most" designs are delivered in 3D, at which time the 2D-to-3D conversion concern might wane.[37] Corroborating evidence from both general market studies,[38] research in peer-reviewed journals,[39] and the evolution of form documents as will be discussed below[40] would suggest the AGC was reasonable in their prediction. The AGC-BIM further offered that the cost of labor for a 2D conversion (assuming a contractor had already made initial investments in software and training) "tends to average between 0.1% and 0.5% of the total construction costs."[41]

In addressing risk management the AGC-BIM states, "much about the legal implications of BIM technology is unknown at present."[42] The guideline notes that while risks presented to contractors by BIM may be different, they are not necessarily greater.[43] For example, contractors will need to understand their ability and right to rely upon a model, but traditional concepts of the architect as the professional in responsible charge of the design should not change.[44] With regards to insurance, the AGC-BIM notes that questions regarding contractor risks associated with BIM use are similar in nature to those questions faced by contractors providing pre-construction services.[45] That is, do certain services a contractor provides in pre-construction – which could result in possible modifications to the architect's design before final completion of the design – increase the contractor's risk? If so, what mitigation steps should be taken? For example, given that commercial general liability insurance responds only to third-party claims for bodily injury and/or property damage, and does not cover purely economic damages, is there the *possibility* that a contractor may want to explore the need for professional liability coverage as well?[46] Additionally, the AGC-BIM notes that the surety underwriting community has, "no clear industry opinion" as of yet with regards to BIM/VDC.[47]

Given the provenance, date of publication, and endorsement of the potential for BIM benefits regardless of project and/or contractor type or size, *The Contractors' Guide to BIM* is a reliable source document in evaluating and determining the evolving standard of workmanlike performance with regards to BIM/VDC. Likewise, as what can be reasonably interpreted and the AGC's continued endorsement of BIM, they now offer a certificate program (Certificate of Management – BIM) which requires completion of both coursework and an exam.[48]

3.3 BIM/VDC source documents – BIM guidelines and standards

Following the general industry research examples noted above, the source documents in this section document continued advancement of BIM/VDC implementation and adoption. Counted amongst the earliest of their kind, these publically available BIM/VDC guidelines and standards reveal universal themes, requirements, and duties regarding BIM/VDC, and hence signal a coalescing of standard of care and workmanlike performance. The source documents reviewed include:

- Indiana University *BIM Guidelines & Standards for Architects, Engineers and Contractors* (2009, Rev. 2012)
- State of Wisconsin *Building Information Modeling (BIM) Guidelines and Standards for Architects and Engineers* (2009, Rev. 9/2012)
- US Department of Veterans Affairs *The VA BIM Guide v1.0* (April 2010)
- *State of Ohio Building Information Modeling Protocol* (April 2010)
- *New York City Department of Design and Construction BIM Guidelines* (July 2012)
- *National Building Information Modeling Standard – United States, v2* (May 2012)

3.3.1 *Indiana University* BIM Guidelines & Standards for Architects, Engineers and Contractors

Indiana University (IU), a state funded research institute, developed initial *BIM Guidelines & Standards for Architects, Engineers and Contractors* in 2009[49] (IU-2009), issuing a revision in 2012[50] (IU-2012). IU-2009 mandated the use of BIM on all projects (new construction, additions/alterations) solicited after October 1, 2009 with total project funding of $5 million or greater, and on any project that was already delivered utilizing BIM. IU-2009 further encouraged, but did not require, BIM on all other projects. IU-2009 issued a parenthetical goal to have all IU projects delivered via BIM by 2011. With the IU-2012 revision, the $5 million requirement threshold remained unchanged with adjustment to the parenthetical goal which now aspires to have all project's using BIM "in the future."[51] The clauses cited and examined below are drawn from IU-2012.

Section 1 Requirements sets forth a number of specific and general BIM requirements. For example, Section 1.1.1, states, in part: "Building information models shall be created to include all geometry, physical characteristics and product data needed to describe the design and construction work of a project. All drawings, schedules, simulations, and services required for assessment, review, bidding and construction shall be extractions from this model." Additionally, the specific requirement of Autodesk Revit as the design team's BIM authoring software and the corresponding .RVT file extension as the format for BIM deliverables are set forth in Section 1.3. Without mandating the use of any particular civil engineering software, additional language in IU-2012 requires the design and construction work to within 5 feet of the building envelope fulfill the same basic requirements as those of the architect/design team.[52] Additional requirements for interoperability,[53] geo-referencing of site plans and building models,[54] and the use of the IU's electronic project collaboration tool[55] (which has the stated feature of a built-in Autodesk Navisworks® viewer) complete Section 1.

Specific procedures for implementing the overall policy of the guidelines are set forth in Section 2 Process. IU-2012 saw the full introduction of a BIM proficiency matrix. Section 2.1 requires the design team, upon request and prior to award, to complete the IU developed BIM proficiency matrix with the stated purpose of enabling IU to get "an overview of the Consultant(s) BIM expertise and experience."[56] Experience with "3D CAD" and SketchUP® are specifically *not* recognized as BIM experience in IU-2012.

Within 30 days of contract award, the design team is required to submit a "BIM Execution Plan." Upon approval by IU (14 days) the BIM execution plan will "be a part of the final bid documents."[57] A review of IU's various publically available standard agreements between the university as owner and an architect[58] (IU-O/A), as well as IU's *Contractor Operating Guidelines*, specifically "Exhibit G Indiana University As-Built And Record Document Deliverables"[59] (IU-COG), show each of those documents to cross-reference the IU-2012. Neither the IU-O/A nor the IU-COG appear to designate a BIM as a contract document per se, but it appears that the aforementioned Section 1.1.1 of IU-2012 requirement that "All drawings, schedules, simulations, and services required for assessment, review, bidding and construction shall be extractions from this model" achieves, essentially, the same.

IU additionally requires the design team to submit within 30 days of contract award an, "IPD (Integrated Project Delivery) Methodology Plan" which upon approval (14 days) will, "be a part of the final bid documents."[60] The IPD plan is intended to demonstrate a high level of design team integration and must include a "critical path methodology on modeling procedures and information validation."[61] Thus, it is important to distinguish between IU's use and meaning of the IPD acronym. IU applies the term in this instance to be specific to design team procedures around modeling, not IPD's more common, recent, acronym as a signifier for separate entities brought together under a single tri-party designer–constructor–owner contract agreement.[62] Section 2 Process contains additional procedures for model quality,[63] and energy requirements.[64]

Section 3 Design Team Deliverable Schedule and Milestones sets forth a basic schedule for various required design team BIM deliverables at each project phase from conceptualization through constructions documents. Section 4 Objective and Applications then details specific requirements (e.g. model content, processes, etc.) by project phase: Pre-design (Section 4.1), Schematic Design (Section 4.2), Design Development (Section 4.3), Construction Documents (Section 4.4), Bidding (Section 4.5), Construction (Section 4.6), Close-out (Section 4.7), and As-built and Record Document Deliverables (Section 4.8). A few noteworthy examples across these sections are highlighted immediately below.

Beginning in schematic design (SD), IU requires the design team to use "automated conflict checking software" to conduct initial collision reporting on the design BIM.[65] Three levels of collision severity are enumerated ("Critical," "Important," and "Important but Recognized to Change") along with a suggested hierarchy of system versus system clash reports to prepare.

At the CDs phase, IU requires the design team to prepare those Construction Operations Building Information Exchange (COBie) design data requirements which are typically the responsibility of the architect.[66] Often aligned with BIM/VDC discussions, COBie, is not concerned with model geometry. COBie is focused on project information, specifically capturing facility asset information for delivery at project close-out and handover.[67] While many BIM authoring tools support extraction of COBie data, the specification itself is built on a Microsoft Excel spreadsheet that can also be manually completed by designers and contractors. Thus, IU-2012 adheres to typical distinctions in COBie responsibilities and workflows and later requires the contractor to prepare their respective portions of the COBie deliverable.[68]

It follows then that definitions and requirements for construction models, in contrast to design models, are set forth in Section 4.6.3. This distinction between design and construction models, and the requirements for each, is significant and most likely reflects IU's intention to leverage any unique and distinct properties of each deliverable type for on-going operations. Whereas as-built construction models (typically produced using 3D CAD as opposed to BIM authoring software[69]) may deliver a portion of useful information related to long-term facilities management, many owners are also interested in conformed design BIMs which can be useful for future architects in future renovation and rehabilitation work. Requirements for contractor collision reporting and concurrent as-builts are detailed in Sections 4.6.5 and 4.6.6, respectively.

Lastly, IU's ownership and rights of data are described in Section 5, reading, in entirety: "Indiana University has ownership of all CAD files, BIM Models, and Facility Data developed for the Project. Indiana University may make use of this data following any deliverable." As one of the first publically available BIM guidelines, many themes and specific elements from IU-2009/12 continue to be referenced and imitated by others throughout the AEC industry.

3.3.2 *State of Wisconsin* Building Information Modeling (BIM) Guidelines and Standards for Architects and Engineers

The State of Wisconsin Department of Administration, through its Division of Facilities Development issued their *Building Information Modeling (BIM) Guidelines and Standards for Architects and Engineers* on July 1, 2009. A revised edition was issued in September 2012 (WI-BIM).[70]

Similar to other state level protocols reviewed in this section, WI-BIM mandates the use of BIM on projects that meet certain public funding minimums. WI-BIM utilizes a threshold trigger system as follows:

- required on all construction (new and addition/alteration) with total project funding of $5 million or greater;
- required on all new construction with total project funding of $2.5 million or greater;

- required on all addition/alteration construction with total project funding of $2.5 million or greater that includes new addition costs of 50 percent or more of total;
- encouraged but not required on all other projects.[71]

Viewed from the perspective of a typical contract stack for design services, the State of Wisconsin's publically available *A/E Contract for Professional Services*[72] references and requires compliance with the State's *Policy and Procedure Manual for Architects/Engineers and Consultants*.[73] The *Policy and Procedure Manual* in turn states, "The Division of Facilities Development established criteria for implementing the use of Building Information Modeling (BIM) on larger projects," and then goes on to enumerate the mandate/dollar value thresholds noted above.[74] Likewise, the *Policy and Procedure Manual* and WI-BIM cross-reference one another.

Neither the *A/E Contract for Professional Services*, *Policy and Procedure Manual for Architects/Engineers and Consultants*, nor the WI-BIM appear to alter typical definitions of "drawings" or "contract documents." Likewise, the *General Conditions of The General Prime Contractor Contract* contains no BIM/VDC-specific definitions or measures and includes typical definitions of "drawings" and "shop drawings." For example: "'DRAWINGS' means the graphic and pictorial portions of the Contract Documents, showing the design, type of construction, location, dimension and character of the Work to be provided by the General Prime Contractor, generally including, but not limited to plans, elevations, sections, details, schedules, diagrams, notes and portions of Specification."[75] However, in describing the BIM requirements for architects and structural engineers the WI-BIM does state:

> Architects and Structural Engineers shall use BIM Authoring software (see section 1.4). Building information models shall be created that include all geometry, physical characteristics and product data needed to describe the design and construction work. *All drawings and schedules required for assessment, review, bidding and construction shall be extractions from this model.*[76] (emphasis added)

The requirement for BIM authoring software is expanded slightly for mechanical, electrical, plumbing, fire protection, civil engineering, and specialty consultants (e.g. food services, security planning, and interior design) with the addition of, ". . . shall use BIM Authoring software *or discipline specialty 3D software*" (emphasis added).[77] For mechanical, electrical, plumbing, fire protection, and civil engineers any specialty 3D software must, nonetheless, "be capable of interfacing with the Architects and Structural Engineers BIM authored software."[78] Similarly, though slightly altered, specialty contractors must be able to interface with only the architect's software.

Despite the expansion allowing the use of non-BIM software (i.e. discipline specialty 3D software) the requirement that all drawings must be extracted from a model is reiterated throughout Section 3 Objectives and Application.[79] Likewise, Section 2.1 Model Quality states, "Do not use disconnected 2D files.

Extract all drawing views from your model." Thus, it might be inferred that on projects which require BIM (per the funding thresholds), BIM is a drawing, and therefore a contract document, insofar as all drawings must be extracted from a BIM and disconnected 2D files are specifically prohibited.

In addition to technical requirements WI-BIM also addresses business processes associated with BIM/VDC. Similar to the State of Ohio BIM protocol, WI-BIM acknowledges the realities of how BIM tools affect typical drawing production rates, schedules, and workflows for architects and their consultants. Accordingly, the State Division of Facilities Development will consider fee allocations which align with "stated BIM work effort limits."[80]

As with the Veteran's Affairs BIM guideline, WI-BIM also has a requirement for Industry Foundation Class (IFC) deliverables at contract close out. This requirement likely signifies their desire to obtain deliverables in a neutral format for any long-term post-construction needs. (A full discussion of IFC is included in Chapter 4.5 Technical – interoperability.)

The WI-BIM reveals the short and long-term goals of public owners to utilize BIM/VDC for controlling project schedule and cost, and prolonging asset lifecycles.

3.3.3 US Department of Veterans Affairs The VA BIM Guide

The US Department of Veterans Affairs (VA) *BIM Guide* (2010) (VAFR-BIM) dates among the earliest federal government tactical BIM guidelines in the United States. While similar in general content structure to other publically available BIM guidelines, the VAFR-BIM is particularly notable for identifying a relationship between the type of governing contract arrangement chosen by the owner and the resulting impact on technical BIM/VDC procedures.

The VAFR-BIM recognizes that the owner's decision to utilize a design–build (DB), or Integrated Design and Construction (IDC) (this term appears original to the VA) contract versus a traditional design–bid–build (DBB) contract arrangement will result in core differences regarding legal concepts such as responsible control and *Spearin* warranties. Those differences will in turn have a direct impact on the BIM/VDC administrative and tactical structures of each respective contract type. For example, Section 2.1 notes that the methods of risk allocation defined in the governing contract will determine "whether there are separate design intent and construction BIM models or, whether they can be combined into one model."[81]

From a mandate perspective, the VAFR-BIM indicates that the policy threshold for BIM use in design and engineering was set for capital projects appropriated at over $10 million (both major construction and renovation) beginning in financial year (FY) 2009.[82] Amongst other sections and provisions, the VAFR-BIM sets forth specific BIM requirements in Section 7 VA Requirements for Using BIM including:

- space and medical equipment validation;[83]
- architecture – spatial and material design models;[84]

- energy analysis;[85]
- COBie/commissioning;[86]
- clash detection/coordination;[87]
- virtual testing and balancing.[88]

Viewed as a contract stack, the VAFR-BIM references the VA's *Program Guide PG 18-15: A/E Submission Instructions for Major New Facilities, Additions & Renovations* (VAPG-AE).[89] That document sets forth the, "minimum submission requirements for the design phase of the major project."[90] The VAPG-AE, in turn, cross-references the VAFR-BIM, as well as explicitly detailing BIM requirements for project phase deliverables.[91]

Stepping back slightly higher in the contract stack, the VA's publicly posted "A/E contracts" incorporate by reference various design standards manuals. For example, the A/E contract for design development (DD) services requires that "VA construction shall be designed in compliance with applicable standards and codes described in Architectural Design Manual PG 18-10."[92] The *Architectural Design Manual* itself then explicitly states that the VA requires "the design team to utilize the BIM tool in designing VA facilities."[93]

So as to not succumb to the tyranny of alpha-numeric contract identifiers, a quick summary of the stack described above: (1) The VA's A/E contracts for various design services incorporate an (2) Architectural Design Manual which requires the use of BIM. Likewise, (3) submittal instructions for design services detail BIM requirements and reference the (4) VAFR-BIM. The VAFR-BIM then sets forth, amongst other provisions, explicit required BIM uses.

Returning to implementation at the project level, the VAFR-BIM requires the development of a "BIM Management Plan" (BMP) to set out the tactical procedures for BIM management and the assignment of BIM responsibilities.[94] Given its aforementioned premise that contract structure drives BIM tactics, the VAFR-BIM requirements call for DB or IDC projects to utilize one BMP to address both design and construction, whereas DBB projects will utilize two separate BMPs for design and construction.[95] In either scenario, the BMP requires measures such as the identification of BIM software tools, as well as risk allocation. Any BMP also requires the parties to establish the legal status of the model. Sections 3.2.o and 3.2.j of respective design and construction BMPs require that models be defined legally as either "Binding, Informational, Reference, Reuse."

Additionally, the BMP requires responsibility assignments for BIM content development and management. To assist in such assignments, as well as to establish and clarify the VA's end-goals, the VAFR-BIM cross-references the VA Object Element Matrix.[96] Self-described as an "expansion of the (AIA) E202"[97] the matrix defines object properties, attributes, and level of development (LOD) requirements throughout the course of design and construction. The object element matrix also serves to support another VAFR-BIM key requirement – commissioning and long-term facility maintenance and operation. As stated in Section 7.8 the "VA has adopted COBie as the methodology to electronically transfer building information after construction is complete for facilities management." As previously noted, the COBie specification offers projects a means of standardizing facility asset information. COBie is not concerned with the geometry of a BIM, but many BIM authoring tools enable their models to extract data for import into the Excel-based COBie worksheet.[98]

Counted amongst the first federal departmental BIM guidelines, the VAFR-BIM likewise represents a guideline document whose themes and specific requirements have been repeated by others.

3.3.4 State of Ohio Building Information Modeling Protocol

The *State of Ohio Building Information Modeling Protocol* (2010) (OH-BIM) presents one of the earliest state-level initiatives in the US to address BIM and VDC. Building upon research and interviews conducted by the State Architect's Office (SAO) the *Ohio Protocol's* stated goals encompass a broad spectrum for BIM including developing a common vocabulary and schema for conveying owner expectations, articulating basic technical requirements, and addressing implementation considerations.[99]

Beginning in July of 2011 the OH-BIM set a dual threshold trigger for determining which state-funded projects would require the use of BIM. First, all projects (whether they be new construction, additions, or alterations) where the total funding through state appropriations is $4 million or greater, require the use of BIM. Or, all projects (again all types) where the total estimated value of plumbing, fire protection, HVAC and electrical work is greater than 40 percent of the value of construction require the use of BIM.[100] These thresholds may also "be adjusted to include any size project or waived by the state agency or institution of higher education for specific project requirements."[101] The OH-BIM further "strongly recommends" BIM use on projects with a construction manager or DB delivery method.[102]

While not specific to the OH-BIM per se, in 2011 the Ohio state legislature also made the first changes to the state's construction procurement methods in over 100 years.[103] House Bill 153 enables multiple state political subdivisions (e.g. municipalities, state colleges and universities, etc.) to, for the first time, utilize contracting methods such as DB or construction manager at-risk with a guaranteed maximum price.[104]

In addition to a BIM requirement rule controlled by public funding and project construction system cost percentage, the OH-BIM addresses the technical realities of how BIM tools and VDC processes affect typical drawing production rates, schedules, and workflows for architects and their consultants. Citing their own SAO research findings that pointed towards "fewer drawing revisions" and "fewer change orders during construction" as a result of the use of BIM, the OH-BIM alters the state's typical payment schedule for architects. Whereas historically architects received 15 percent of fee disbursement for SD, 15 percent for DD and 30 percent during the production of CDs, Ohio now disburses 20 percent at each of those phases – SD, DD, and CDs – on projects implementing BIM. Additionally, the OH-BIM Protocol states, "The use of BIM on a project should not result in increased fees."[105]

The OH-BIM also provides examples of preliminary contract provisions that "have been, or will be included, in the suite of legal documents that frame the contractual framework for each project."[106] Examples (presumably noting standard SAO contract sections) include language addressing, introductory team consensus building,[107] conformed documents,[108] document ownership and use,[109] architect/engineer intellectual property,[110] and indemnification for use of electronic files.[111] With regards to this last example concerning indemnification for use and reliance, sub-section .8 reads, "In the event of a conflict between the Contract Documents and the Electronic Files, the Contract Documents shall control, take precedence over, and govern the Electronic Files."[112]

As an example of an early state-level BIM guideline, many aspects of the OH-BIM have been borrowed throughout the AEC industry. Amongst other features, the OH-BIM is notable for being among the first guidelines to alter fee schedules for architects to better align with the BIM/VDC workflows that alter 2D drawing production through extraction.

3.3.5 New York City Department of Design and Construction BIM Guidelines

The New York City Department of Design and Construction (DCC), Public Buildings Division is charged with providing design and construction project management services for an array of municipal agencies within the city government. The DDC portfolio of projects, valued at nearly $10 billion, involves both new and renovation projects across a broad spectrum of building typologies: libraries, police precincts, courts, public parks buildings, transportation facilities, and so on. The DDC published its *BIM Guidelines* (DDC-BIM) in July 2012.[113]

Before examining the DDC-BIM itself, a brief overview of the DDC's contracting structure. The DDC serves as the contract holder with a given project's design, engineering, and construction professionals on behalf of their (DDC's) municipal agency clients. The DDC's Public Building's design work is issued under two contract groupings known as the "6/20." One group consists of six (6) firms hired to work on projects where the construction value is between $15 million and $50 million. A second group consists of twenty (20) small firms who are hired to work on projects up to $15 million in construction value.[114] Recent (2012) publicly available DDC requests for proposals (RFPs) relative to the groups of "6" and "20" are reviewed below to better comprehend the DDC-BIM within a typical design contract stack.

RFP-6, Six Architectural and Engineering Design Requirements Contracts for Large Projects, Citywide (RFP-6) references and includes an attached requirements contract, *Requirements Contract for Architectural, Engineering and Construction Related Services* (DDC-6 Req. Contr.).[115] The DDC-6 Req. includes, amongst its named exhibits, *Design Consultant Guide* (DDC-CG) and the DDC-BIM. The DDC-6 Req. neither alters typical definitions for "drawings," or "shop drawings" with respect to BIM, nor includes new definitions relative to BIM/VDC. Its definition for "Contract Documents" does, however, include the DDC-BIM as an exhibit.[116]

The DDC-6 Req. also specifies that all required design services "shall be in accordance with . . . (4) Guide for Building Information Modeling (BIM) Services"[117] and states that BIM services are to be included within the calculated design fee.[118] The DDC-6 Req. specifically requires the architect of record to affix his stamp to permit drawings.[119] The DDC-BIM affirms the primacy of 2D contract sets of drawings.

> When conflicts exist between the contents of a BIM and the Contract Set of Drawings, the information contained within the Contract Set will prevail and will be considered as definitive. The BIM shall still contain accurate representations of the design condition regardless of what is displayed on the drawing set.[120]

The contemporaneous RFP and associated requirements contract for the group of "20," *RFP-20, Twenty Architectural & Engineering Design Requirements Contracts for Small Projects, Citywide* (RFP-20), make reference to the same general design guidelines (DDC-CG); however, neither references the DDC-BIM, nor makes any independent mention of BIM. The introductory summary to RFP-20 notes:

> DDC is limiting the size of the firm that is eligible for these contracts to those with no more than ten (10) professional staff. ("Professional Staff" includes all design, production and construction staff from principal to junior draftsperson – this also includes any other professional disciplines such as interior design, engineers or landscape architects).[121]

Thus, from a strict contract perspective of the examples reviewed, the DDC appears to have set the threshold for BIM requirements by both dollar value of construction (> $15 million) and firm size (> ten professional staff).

It also appears, however, that BIM could be deemed a requirement on smaller projects. For example, as the general design guideline strategy set forth in the DDC-CG notes: "DDC considers that Building Information Modeling (BIM), as both technology and process, is superior to traditional non-BIM methods, when properly scaled in its use . . . In addition to this guide, for all BIM designated projects, consultants shall adhere to the latest version of the DDC BIM Guidelines."[122] Likewise, in describing the scope preparation and design consultant selection process the DDC-CG indicates that BIM applicability will be determined at the outset of each project, during the first of four project stages.[123]

The DDC-BIM itself is organized into three primary sections: Part One: General Information, Part Two: BIM Uses and Requirements, and Part Three: Submission and Deliverables. Included in Part One are statements concerning the general purpose of the DDC-BIM and the Public Buildings Division complete lifecycle vision regarding BIM. Part One also includes an inclusive list of currently known and acceptable BIM software. While as discussed above the DDC requires 2D permit drawings to be stamped by the architect of record, the DDC-BIM does identify specific BIM software that is capable of "Code Checking."[124]

Part One also sets forth specific requirements for named BIM roles and respon-sibilities, and the requirement for a "BIM Execution Plan" (BEP) on each pro-ject. As a tactical and operational document the BEP is required to detail how a project will be "executed, monitored and controlled with regard to BIM."[125] The BEP must be completed within 30 days of a project's contract registration, and align with the project's chosen delivery method. Similar to other BIM guide-lines reviewed in this section, the DDC-BIM notes that the BEP structure will be impacted by delivery methods, that is, whereas a DBB project will have two BEPs – one for design, one for construction – an IPD may suffice with one BEP for the project's duration.[126] Additionally, Part One addresses model ownership, stating in part:

> DDC holds ownership of the BIMs including all inventions, ideas, designs, and methods contained within the model. This includes, but is not limited to; the content submitted as part of the BIMs itself. Outside resources, such as consultants and/or contractors, using the BIM are granted temporary use of it for the duration of the project.

Part Two: BIM Uses and Requirements sets forth descriptions of common BIM use applications that a project might employ. Any chosen use applications on a given project are then to be documented in the BEP. The use applications span the complete project lifecycle, and include: existing conditions modeling (which sub-references laser scanning and the DDC's "Site Engineering Laser Scanning Group"), sustainability (LEED) evaluation, cost estimation, record modeling, and asset management. Of particular note is the use application for "Code Validation." The DDC-BIM states the value of utilizing BIM for code vali-dation to include, "Validate that building design is in compliance with specific codes (IBC International Building Code, ADA Americans with Disabilities Act guidelines and other project related codes."[127]

Part Two also outlines LOD definitions. The DDC-BIM explicitly states that the LOD 100-500 definitions are based on the AIA E-202 (2008).[128] To assist users of the document, the DDC-BIM includes graphic exemplars of each LOD.[129] Additionally, an explanation of "Model Granularity" is provided:

> BIM's shall be created providing an accurate representation of geometry needed to support specific BIM use. The level of detail needed will vary by object and by model, and the BIM itself may not represent the exact design intent of real live elements. As a rule of thumb, any object that fits within a 6"×6"×6" cube should not be modeled.[130]

Part Three: Submission and Deliverables notes that at each design and construction phase the DDC requires submission of "the model, electronic versions of hardcopy submissions and other files that support the intent of the project." Part Three then proceeds to provide BIM submission requirements and details for each project phase, from pre-schematic through construction.

Phase-specific details include, for example: software file extension types (e.g. .RVT, .NWD), cost estimation requirements, clash detection hierarchies, construction scheduling mandates, and asset management object data fields. Part Three ends with a review of the DDC's required naming conventions and standards for BIM projects, including: file naming, discipline coding, and BIM object naming.

As one of the largest and most complex cities and municipal governments in the world, the development of a BIM guideline by the New York City DDC suggests that that organization recognizes value in the utilization of BIM/VDC. Indeed, the DDC-BIM is clear that it considers BIM/VDC, when properly implemented and scaled, to be far superior to traditional technologies and processes. As with other publically available BIM guidelines the themes and specific requirements of the DDC-BIM have been copied and repeated by others in the AEC industry.

3.3.6 National Building Information Modeling Standard – United States, v2

In what might be reasonably interpreted as a direct response to the identified deficiencies and AEC industry call-to-arms of the NIST-InOp report reviewed above, the buildingSMART™ alliance, under the umbrella of the National Institute of Building Sciences released the *National Building Information Modeling Standard - v2* (NBIMS-v2) in 2012. The NBIMS-v2 stated objective is to "advance the art and science of the entire life-cycle of the vertical and horizontal built environment by providing a means of organizing and classifying electronic object data."[131] The scope of the NBIMS-v2 was developed specifically for two distinct groups: (1) software developers and vendors, and (2) design, engineering, construction, and operations professionals.[132]

At more than 600 pages (inclusive of appendices), the NBIMs-v2 is a voluminous document. Accordingly, a complete in-depth analysis of the entire document is beyond the immediate scope of this section. Likewise, while not to suggest in any way that those sections focused on the software developer (i.e. reference standards and exchange information standards) are less relevant than the practice documents section for professionals, for the objectives herein (investigation into BIM/VDC standard of care and workmanlike performance at the end-user level) the latter section is selectively reviewed. Furthermore, having now chosen the practice documents sections as the particular focus of analysis, it must be noted that NBIMS-v2 has self-assessed its own professional practice document sections as being, "Thus far in the NBIMS development process . . . the least documented for the building disciplines."[133]

The practice documents section begins by informing readers that the NBIMS-v2 is "by design, a standard of standards."[134] Thus, the reader is presented with the reality that, dependent upon their specific needs, any standard included within NMBIS-v2 may be selected, but that, essentially, the precise tactical means and methods used to achieve those needs are ultimately the purview of the contracting or interested parties.[135] What this means practically, and without judgment, is that the NBIMS-v2 presents a competent overview of the current state of BIM practice processes and documents, but that it does not represent a specific codification of the "rules and regulations" of BIM as an outside reader to that document might initially anticipate from its title. Again, this does not suggest NBIMS-v2 impotence, but for the stated analysis of the present section, the meaning of "standard" within "Standard" must be noted.

For example, Chapter 5.5 is titled, "Mechanical, Electrical, Plumbing, and Fire Protection [MEPFP] Systems Spatial Coordination Requirements for Construction Installation Models and Deliverables – November 2009, Revised May 2012." As the title implies, the section offers guidance in the spatial coordination of MEPFP systems and components, but the chapter's introduction is unequivocally clear that "It is not a replacement for the ConsesusDOCS 301 *BIM Addendum*, a BIM Implementation Plan, or any other more project specific scope of work or contract."[136] Again, at the expense of repetition, and offering no judgment with respect to any part of the NBIMS-v2, the current analysis simply demands a reader to contemplate "Requirements" in the chapter heading within the overall work's title.

With "standard" and "requirements" under the microscope, it does not seem unreasonable to imagine that most AEC industry professionals in the US approach the NBIM-v2 with at least some degree of direct experience with the United States National CAD Standard[137] (NCADS), and perhaps therein lies a quandary. In its fifth version (as of this research), the NCADS provides a developed coordination of three distinct standards from three distinct stakeholder groups or perspectives, namely: The American Institute of Architect's *CAD Layer Guidelines*, the Construction Specification Institute's *Uniform Drawing System* (Modules 1–8), and the National Institute of Building Sciences *Plotting Guidelines*.[138] Thus, the NCADS integrates tactical documents that are primarily focused on techniques and methods for organizing static digital/ink on the screen/page – layer naming guidelines, schedule components, graphic symbolism, and so on. As numerous scholars of the AEC industry have noted, the advent of 2D CAD essentially digitized the mechanical process of drafting by hand with ink or pencil.[139] Thus, the then burgeoning CAD standards movement would necessarily have had as its primary objective defining new methods for delivering age-old and familiar 2D contract drawings. In contrast, BIM/VDC are changing not only the rote method of production for contract drawings, but the complete spectrum of processes associated with design and construction. Indeed, it would appear fair to suggest the existence of the NBIMS-v2 itself begins to open a longer discussion around definitions of contract drawings in the first instance.

3.4 BIM/VDC source documents – form contracts

The BIM/VDC source documents selected for review in the two preceding sections provided an overview of general industry research, followed by industry guidelines and standards evidencing the establishment of standard of care and workmanlike performance. This section will review recent and current

BIM/VDC-specific form contract documents by the AIA and ConsensusDocs. These form contract documents further clarify current typical BIM/VDC requirements, duties and expectations, and formalize them in recognizable, industry-accepted formats. The source documents reviewed include (see Appendix A):

- AIA – E202™ (2008) *Building Information Modeling Protocol Exhibit*
- AIA – E203™ (2013) *Building Information Modeling and Digital Data Exhibit*
- AIA – G201™ (2013) *Project Digital Data Protocol*
- AIA – G202™ (2013) *Project Building Information Modeling Protocol Form*
- ConsensusDocs – 301™ *Building Information Modeling (BIM) Addendum*

3.4.1 AIA – E202™ (2008) Building Information Modeling Protocol Exhibit

In the introduction to his thought provoking work *The Death of Contract*, the late legal scholar Grant Gilmore noted that even the most intensely detailed observation of the present state of things only becomes useful "when we are in a position to compare it with what we know about what was going on last year and the year before that and so on back through the floating mists of time."[140] Thus instructed, before attempting to meaningfully observe the present state of things through the AIA E202™ (2008) *Building Information Modeling Protocol Exhibit*, and its current successor documents, E203™ (2013), G201™ (2013), and G202™ (2013), a short historical context is offered.

The AIA typically exhibits a 10-year cycle for revision and re-write of their form documents.[141] However, with respect to contract documents categorized as "Digital Practice" the AIA has issued a total of seven form documents (across three series: C, E, and G[142]) over a 6-year span (2007–2013). Specifically, the E202™ *Building Information Modeling Protocol Exhibit*, first issued in 2008 and counted amongst the inaugural industry form documents explicitly tailored for projects utilizing BIM, was revised within a 5-year cycle. In addition to its brisk revision pace, the E202™ released in 2008 as a singular document was by its 2013 revision re-configured and expanded as three separate documents, namely, E203 (2013), G201 (2013), and G202 (2013), to address both BIM specifically, and project digital data generally.

1   C106 (2007) *Digital Data Licensing Agreement (RETIRED)*
2   C106 (2013) *Digital Data Licensing Agreement*
3   E201 (2007) *Digital Data Protocol Exhibit*
4   E202 (2008) *Building Information Modeling Protocol Exhibit*
5   E203 (2013) *Building Information Modeling and Digital Data Exhibit*
6   G201 (2013) *Project Digital Data Protocol Form*
7   G202 (2013) *Project Building Information Modeling Protocol Form*

As others have noted, the non-profit AIA produces new documents when it feels there is a compelling and pressing market demand for one.[143] The AIA's current pace of production for BIM and VDC specific documents within a shorter than typical span would appear to be filling such market needs. Granted, existence of such form documents may not necessarily be indicative of widespread awareness and use of such documents by a majority of stakeholders across all geographies.[144] Conversely, lack of formal contractual arrangements for BIM and VDC on a given project may not be indicative of actual BIM and VDC usage on a project.[145] Notwithstanding the possibility of either scenario, for an analogous historical perspective on robust revision pace consider the AIA's revision cycle for BIM and VDC documents with their revision of the A201™ *General Terms and Conditions* during the so-called liability explosion beginning in the 1960s. The period between 1961 and 1982 saw no fewer than seven iterations of the *General Terms and Conditions*.[146] While not suggesting mirror fact patterns or market influences or conditions, both trends might be safely categorized as real and significant pressures tangibly altering the daily practice of design and construction.

In conjunction with this general summary and chronological review of the AIA's digital practice documents, additional context for how the AIA has addressed digital technology can be traced through their handbook for professionals. For example, the 14th edition of the AIA's *The Architect's Handbook of Professional Practice* (2008) was the first edition to specifically address BIM substantively.[147] While the earlier 13th edition (2001) noted that parametric modeling was "poised to supplant vector-based CAD in building design software,"[148] the 14th edition is the first to specifically index and treat "building information modeling." Indicative of the pace of technological transformation in architectural practice, the 12th edition of the *Handbook*, published in 1994, provided practitioners with a comprehensive overview of 2D CAD systems, including, amongst other items, a detailed section dedicated specifically to "CAD Layer Guidelines."[149] From a parallel review of contemporaneous AIA form documents, it appears that the 1997 version of the B141™ *Agreement Between Owner and Architect* (replacing the 1987 edition) was the first AIA contract document to explicitly note "electronic modeling" as an example of a potential deliverable during the schematic phase of design.[150] Stepping back only slightly further on the timeline, the 11th edition of the *Handbook* published in 1988 was the first to address CAD at all, and lightly at that.[151] The 10th edition, published in 1970, noted advantages to preparing CDs using pencil on tracing cloth.[152]

Against such a backdrop unfolds an analysis of the E202™ (2008) *Building Information Modeling Protocol Exhibit*. Structurally, the E202™ (2008) is divided into four articles:

1 General Provisions
2 Protocol
3 Level of Development
4 Model Elements.

As the title implies, the E202™ (2008) is not a stand-alone document. Article 1 includes a provision that it must be attached as an exhibit to any other executed agreements for services or construction on the project.[153] While requiring that the exhibit be attached to any other executed agreement, the E202™ (2008) does not set forth a specific time requirement for doing so. Thus, in practice, a project's governing contracts could be executed well before the execution of the exhibit, or before BIM was even thoroughly considered a part of the delivery process amongst all a project's stakeholders. Such timing left many early BIM projects in the US contractually toothless, insofar as late-addition E202™ (2008) exhibits had BIM "requirements" that often provided more owner appeasement than realized value. Accordingly, as will be discussed in due course below, the AIA's subsequent revision and restructuring of the E202™ (2008) into the E203™ (2013), G201™ (2013), and G202™ (2013) provides for essential BIM discussions to occur at the beginning of a project. Timing considerations aside, Article 1 also states that where a conflict exists between a provision in the E202™ and any executed Agreement, the provision in the exhibit will take precedence.[154]

Article 1 next provides definitions for key terms. Within the exhibit, "Building Information Model" describes the physical and functional characteristics of a project, and may be referred to as, "the "Model(s), which term may be used herein to describe a Model Element, a single Model or multiple Models used in the aggregate."[155] The definition for "Model Element" is subsequently defined as a distinct portion of a BIM, "representing a component or assembly within a building or building site . . . organized by the Construction Specifications Institute (CSI) UniFormat™ classification system."[156]

Accordingly, within the E202™ the term "Model" might be used interchangeably to refer to: a mechanical design engineer's complete HVAC design model, an aggregated model of all design consultant models as compiled by an architect, or a specific shop model of an HVAC distribution system on a given floor of a multi-story building as produced by a subcontractor. Thus, at a root level, no dedicated terms are provided for distinction between models produced for design phases or activities, versus models produced for construction phases or activities. However, in later defining the "Model Element Author" (MEA) as the "party responsible for developing the content of a specific Model Element" whose responsibility is then subsequently documented in the Section 4.3 Model Element Table, the E202™ does enable modeling responsibilities between the architect and contractor to be identified and clearly distinguished. Also included in Article 1 are definitions for "Model User" which refers to any individual or entity authorized to use the model, and "Level of Development."

While any other term defined in Article 1 might not be easily recalled by a US AEC professional conversant with BIM, the term level of development, or "LOD," has arguably entered the vernacular of US BIM practice with as much force as the acronym BIM itself. As defined in the E202™, LOD "describes the level of completeness to which a Model Element is developed" and provides for the now ubiquitous (amongst BIM conversations) five levels of progressive model element completeness – LOD 100 through LOD 500. The specific definitions of each LOD are defined in Article 3 and will be reviewed in due course below after examining Article 2.

Article 2 begins by setting forth duties regarding conduct, establishing rules of ownership, and providing for the establishment of macro-level procedures. With respect to conduct, Section 2.1 Coordination and Conflicts requires that "regardless of the Phase of the Project or LOD" where conflicts are found in the Model the discovering party must promptly notify the responsible MEA(s) who then must promptly act to "mitigate the conflict." Being a BIM-specific Exhibit, "conflict" is most likely understood by project participants to mean issues dealing with spatial coordination of systems. While not suggesting parity between the two terms, this now technologically-driven duty to reveal "conflict" would appear to relate – at least abstractly – to the established construction law concept of a contractor's "duty to inquire." In light of this, a brief synopsis of several cases concerning this concept.

In *Merando, Inc.* v. *United States* the court held that where a contractor is faced with an obvious discrepancy in a set of bid drawings, the contractor must inquire and seek clarification before proceeding if he wishes to benefit from the discrepancy in his bid.[157] As a bidding contractor, Merando had received an initial invitation to bid along with associated bid drawings, and then subsequent revisions and amendments to those bid drawings. The court noted that each set of drawings, "contained the usual 'contract limit line' within which the contract work was to be performed."[158]

During the course of design revisions and amendments, an irregular circle had been specifically drawn to highlight and call attention to additional work to be included as part of the contract, but which lay outside the originally drawn contract limit line. Merando's bid disregarded the additional work highlighted by the irregular circle. Post-award, Merando subsequently requested compensation for the additional work outside the contract limit line. The court noted that, ideally, the contract limit line should have been moved to encompass the new work, but held, "Faced with the patently obvious discrepancy between the placement of the contract limit line and the additional work, plaintiff should have inquired. This pre-bid inquiry would have clarified the situation and allowed the contractor to bid in accordance with the requirements of the contract."

Similarly, in considering interpretation of two ambiguous structural reinforcing details in a drawing set, the court in *Fortec Constructors* v. *United States* held,

> The existence of a patent ambiguity in the contract raises the duty of inquiry . . . [citation omitted] . . . In deciding the correct meaning of a contract containing a patent ambiguity it is proper to consider the trade standards and practices of the relevant business community [citation omitted]. A contractor is charged with the knowledge of these standards.[159]

Each of the cases above deal with vagueness in printed drawings where the degree of uncertainty shown on the drawing is easily recognizable or obvious. Each case contains patent ambiguity at a different scale: *Merando* at the macro level of an entire scope of work, that is, "what work is included in the contract where the drawings show greater context?" and *Fortec* at the level of a structural detail, that is, "of multiple solutions shown, which one must be used?" Neither deals with the discovery and resolution of "conflict" of spatial coordination as anticipated by the duty set forth in the E202™. Nonetheless, retroactively considering BIM technologies in either scenario opens forward thinking discussion.

For example, inherent capabilities of BIM technologies reach beyond simply identifying where model element "A" intersects with model element "B" in space. If BIM parameters[160] – such as "contract," and "not in contract" – were applied to each model element in a BIM, would a designer and bidder such as those in *Merando* then be able to more easily clarify visually their intention and expectation, respectively? Or, given existing capabilities of structural-specific BIM technologies, would the ability to visualize ambiguous rebar details such as those in *Fortec* resolve the ambiguity, per se?

Having set forth the duties to notify and act with respect to conflicts, Article 2 then establishes that data ownership does not extend beyond the MEA. Section 2.2 Model Ownership states:

> In contributing to the Model, the Model Element Author does not convey any ownership right in the content provided or in the software used to generate the content. Unless otherwise granted in a separate license, any subsequent Model Element Author's and Model User's right to use, modify, or further transmit the Model is specifically limited to the design and construction of the Project, and nothing contained in this exhibit conveys any other right to use the Model for any other purpose.

Given owners interest in using BIM deliverables post-construction for facilities maintenance and operations requirements,[161] any "separate license" as noted above is worth consideration. For example, if used in conjunction with the AIA B101™ (2007) *Standard Form of Agreement Between Owner and Architect*, Article 7, Section 7.3 of that document grants to the owner a nonexclusive license to use the architect's "Instruments of Service solely and exclusively for purposes of constructing, using, maintaining, altering and adding to the Project, provided that the Owner substantially performs its obligations." More directly, many owner-driven BIM standards have explicitly addressed the issue of BIM ownership to their advantage. For example, as previously noted, the IU BIM *Guidelines* state, "Indiana University has ownership of all CAD files, BIM Models, and Facility Data developed for the Project. Indiana University may make use of this data following any deliverable."[162] Similarly, the *New York City Department of Design and Construction BIM Guidelines* state, in part, "DDC holds ownership of the BIMs including all inventions, ideas, designs, and methods contained within the model. This includes, but is not limited to; the content submitted as part of the BIMs itself."[163]

Article 2 continues on to provide for the establishment of macro-level procedures within Section 2.3 Model Requirements. Parties are required to establish if a model standard will be utilized. While not prescribing the use of any particular model standard, or requiring any at all, Section 2.3.1 parenthetically notes the *National Building Information Modeling Standard*[164] as a possible model standard. Section 2.3.2 then requires a determination of which file format(s) are required as appropriate to particular model uses.

Having provided for macro-level protocol considerations, Article 2 segues to establishing more specific responsibilities, tactics and procedures. Section 2.4 Model Management requires project participants to identify the party responsible for model management by each project phase. As an AIA authored document, the E202™ defaults to the architect as the party responsible for model management from the inception of the project. Once the model manager by phase is identified, detailed tactical responsibilities are enumerated. Section 2.4.2 Initial Responsibilities includes requirements such as establishing a model origin and coordinate system, a file storage protocol, and clash detection protocol(s). Section 2.4.3 Ongoing Responsibilities addresses more iterative processes, such as: coordination and submission of models, maintaining model archives and backups, and managing access rights.

Lastly, Article 2 establishes responsibilities and protocols for model archives. Section 2.4.4 Model Archives states, in part, "The Party responsible for Model management . . . shall produce a Model Archive at the end of each Project phase and shall preserve the Model Archive as a record that may not be altered for any reason." Section 2.4.4.1 then indicates that the model archive shall consist of two sets of files – the first as individual models in their native file format and condition, and the second set as the aggregate of those models, "suitable for archiving and viewing."

Article 3 provides definitions for each LOD (100 to 500). As noted above, the E202™ defines LOD as "the level of completeness to which a Model Element is developed."[165] LODs are cumulative, such that LOD 300 includes and builds upon the characteristics of LODs 200 and 100.[166] Conceptually LODs were initially interpreted by the industry to correspond to a progressive project phase, that is, SD (100) through as-built construction (500). For example, the definition for LOD 300 (roughly analogous to CDs) reads, "Model Elements are modeled as specific assemblies accurate in terms of quantity, size, shape, location and orientation. Non-geometric information may also be attached to Model Elements." However, the realities of the design, construction, and modeling process can result in variable LODs per object or system within a BIM at a given point in time. As such, a BIM is never collectively, say, a "LOD 200 BIM." This initial perception has been self-correcting in the industry and subsequent industry guidelines such as the BIMForum's *LOD Specification* seek to further clarify best-practice LOD usage.[167]

In addition to defining the geometric and spatial characteristics of model elements at each phase, each LOD identifies and defines various "Authorized Uses" of model content at the given LOD. Authorized uses seek to provide MEAs the ability to convey and control expectations and use of the content by others. For example, the authorized use for cost estimating at LOD 200 states, "The model may be used to develop cost estimates on the approximate data provided and conceptual estimating techniques. (e.g. volume and quantity of elements or type of system selected.)"[168]

As the final article of the E202™, Article 4 contains a model element table which is meant to bring cohesion and functional structure to the trio of model elements organized via the Uniformat® classification system, LOD definitions, and respective MEAs. The table provides a blank matrix for project participants to assign a responsible MEA and LOD for each building element or system at the conclusion of any given project phase. In light of the note above about variable LOD status within a BIM at any given time interval, the use of the word "phase" in Section 4.3 was likely a contributing factor to early confusion. As recast is the successor G202, the model element table instructions now replace "phase" with "milestone."[169] Accordingly, the model element table provides a simplified visual reference for project participants to understand who is modeling what, to what level of completeness they are going to do so, and what rights other users of their data have in utilizing, interpreting, and acting upon that data.

Article 4 also contains language clarifying the purpose, expectations, and rules for reliance on model elements as identified in the model element table. Section 4.1.1 dictates that model elements are understood to be shared with other MEAs or model users throughout the project. In recognizing that BIM content creation by an MEA may result in certain content containing more information than required for a given LOD at a given phase, Section 4.1.2 states, in relevant part, "Model Users and Model Element Authors may rely on the accuracy and completeness of a Model Element consistent only with the content required for the LOD identified in Section 4.3 [the Model Element Table]." For example, one might envision a scenario in which a mechanical design engineer has downloaded BIM content from a specific equipment manufacturer for a particular type of pump during early phase DD. While the content as developed by the manufacturer may contain a high level of geometric and graphical informational detail (e.g. performance specifications, etc.), the rules of engagement dictate that it can only be utilized and relied upon by others consistent with the LOD assigned for the given phase. Furthermore, Section 4.1.3 provides:

> Any use of, or reliance on, a Model Element consistent with the LOD indicated in Section 4.3 by subsequent Model Element Authors or Model Users shall be at their sole risk and without liability to the Model Element Author. To the fullest extent permitted by law, subsequent Model Element Authors and Model Users shall indemnify and defend the Model Element Author from and against all claims arising from or related to the subsequent Model Element Author's or Model User's modification to, or unauthorized use of, the Model Element Author's content.

With a summary review of the inaugural E202™ (2008) now complete, an analysis of the present state of the AIA's BIM documents is perhaps enriched.

3.4.2 AIA – E203™ (2013) Building Information Modeling and Digital Data Exhibit

Collectively, the E203™ (2013), G201™ (2013), and G202™ (2013) replace and restructure the E202™ (2008). At the broadest stroke, the updated E202™ (2008) is now recast as an exhibit – the E203™ (2013) – and two protocols – the G201™ (2013) and G202™ (2013). In a significant departure from its predecessor, the E203™ (2012) provides a mechanism for establishing general intentions regarding both BIM and digital data at the time governing agreements are executed. As related protocols, the G201™ (2013) [Digital Data] and G202™ (2013) [BIM] then offer means to provide more specific requirements, responsibilities, and duties at more applicable intervals in the project's contractual timeline. This separation of overall goals from explicit details stands in contrast to the single document format of the E202™ (2008) which required parties to simultaneously establish their general intentions *and* provide tactical protocols regarding BIM in one fell swoop, often long after governing agreements had already been executed, or before all project participants were included.

Organizationally, the E203™ (2013) encompasses five articles:

Article 1 General Provisions

Article 2 Transmission and Ownership of Digital Data

Article 3 Digital Data Protocols

Article 4 Building Information Modeling Protocols

Article 5 Other Terms and Conditions.

As the title reveals, the E203™ addresses both BIM and digital data. Section 1.1 states the general purpose of the exhibit to document expectations around each domain. The definition for BIM (inclusive of model) is essentially the same as that provided in the E202™ (2008). Newly added "Digital Data" is defined as "Information, including communications, drawings, specifications and designs, created or stored for the Project in digital form. Unless otherwise stated, the term Digital Data includes the Model."[170] Practically, amongst other possible benefits, the distinction and inclusion of model within the broader term digital data enables the contracting parties to specifically identify digital, *non-model* information (e.g. submittals, payment documentation), and actively consider and indicate where any such digital non-model information may or may not intersect/interact with a model. For example, a number of project and document management solutions allow for two-way interaction between 2D graphical (e.g. 2D floor plans) or text information (e.g. text-based specifications) and a Model.[171]

In another departure from its predecessor, the E203™ adjusts the process for incorporation of the exhibit into other project agreements. The predecessor E202™ (2008) was developed as an exhibit attached to a specific agreement identified on its title page, which was then incorporated by reference into other project agreements for services or construction.[172] In contrast, the E203™ (2013) is struc-tured to be the single, global initiating source of BIM expectations for the project and obliges contracting parties to incorporate it into any other project agreements whose project participants will utilize digital data on the project. In line with this flow-down provision, the E203™ separately defines parties (signatories to an agreement) and project participants (individuals or entities providing services, work, equipment, or materials). Project participants also includes parties.[173]

Further, parties – as providers of digital data – may require other project par-ticipants – as requestors – to provide "reasonable evidence" that the exhibit and most recent versions of the tactical protocols (G201™ (2013) [Digital Data] and G202™ (2013) [BIM]) have been properly incorporated into their (the request-ors) project agreements before transmitting or giving access to digital data.[174] In short, if you want it, prove your entitlement.

Additionally, the parties also explicitly agree via Section 1.2.1 that any project participant utilizing digital Data is an intended third-party beneficiary of the obliga-tion to include the exhibit into other project agreements. For example, if a vendor for an air handling unit had a direct agreement with a contractor that was required to, but did not, incorporate the E203™ and the vendor's engineers then improperly utilized the architect's LOD 200 model elements to prepare fabrication drawings resulting in a claim, the architect could protect himself by bringing a breach of contract claim against the contractor for failure to incorporate the E203™.

Given that the E203™ is intended to be executed at the onset of a project, Article 1 further establishes procedures for requesting and negotiating any addi-tional compensation, contract sum, schedule or contract time that may arise from the requirements and responsibilities set forth in the exhibit and associated pro-tocols.[175] As previously noted, the predecessor E202™ (2008) was often prepared long after any governing agreements had been executed and project budgets and schedules had been set. As a result, some projects progressed to a point at which BIM was more seriously considered, only to have requests for adjustment to com-pensation or time to meet any emerging BIM requirements fall upon deaf ears. While not suggesting the revised E203™ guarantees access to time or money, it nonetheless requires project participants to consider impacts associated with BIM "from Day 1," thereby increasing the chances that expectations are clearly aligned with financial realities.

Procedurally, unless otherwise provided, the party seeking any such adjustment must notify the other party within 30 days of receipt of the protocols.[176] Upon notification the parties are required to "discuss and negotiate" any changes to the protocols, or changes to the requested monies or time within the terms of the governing agreement.[177] An example of a changes clause from a governing agree-ment might be akin to Article 8 of the ConsensusDocs 200 *Standard Agreement and General Conditions Between Owner and Contractor* which reads, in part:

> 8.1.1 The Contractor may request or the Owner may order changes in the Work or the timing or sequencing of the Work that impacts the Contract Price or the Contract Time. All such changes in the Work that affect

Contract Time or Contract Price shall be formalized in a Change Order. Any such requests for a change in the Contract Price or the Contract Time shall be processed in accordance with this Article 8.

Article 1 concludes by providing definitions for new terms, and refining/clarifying terms previously defined in the E202™ (2008). For example, the definition for LOD now states, "(LOD) describes the minimal dimensional, spatial, quantitative, and other data included in a Model Element to support the Authorized Uses associated with such LOD."[178] Additional new terms include: confidential digital data,[179] written or in writing,[180] and written notice.[181]

Rights of data ownership in the E203™ (2013) are dealt with in fundamentally the same manner as in the E202™ (2008) – that is, a party transmitting digital data conveys no ownership rights to the receiving party. Ownership rights are controlled by the governing agreements.[182] Similarly, while the rights of any party receiving data are limited to using, modifying, and further transmitting the data only as it relates to the project, the E203™ (2013) does expand applicable uses beyond just designing and constructing the project. In a likely response to commentary by owners wanting to use BIM deliverables for operations and maintenance post-construction, Section 2.3 expands usage rights beyond design and construction to include, "using, maintaining, altering and adding to the Project." This language matches that found in Section 7.3 of the AIA's B101™ *Owner Architect Agreement (2007)* that provides owners with a nonexclusive license to use an architect's instruments of service.

Along with data ownership, Article 2 provides clauses controlling transmission of digital data. A party transmitting digital data is required to warrant that it owns the copyright on the data it is transmitting, "or otherwise has permission to transmit the Digital Data for its use on the Project."[183] Separate clauses, Section 2.2 and Section 2.2.1, establish similar transmission duties with respect to "Confidential Digital Data" which is defined as, "Digital Data containing confidential or business proprietary information" that has been clearly marked as such.[184]

Having outlined the general provisions of the exhibit (Article 1), and set forth the specific duties and rights with respect to digital data ownership and transmission (Article 2), the E203™ then embarks more directly on its stated mission to establish a basic level-set amongst project participants regarding Digital Data (Article 3) and Building Information Modeling (Article 4).

Given that the E203™ is intended to be executed at the same time as the governing agreements, Article 3 allows the parties to establish their basic assumptions, intentions, and anticipated types of digital data as they then exist. The specific requirements and tactics will subsequently be set forth in the protocol document, AIA–G201™ (2013) *Project Digital Data Protocol*. As has been noted above, Section 1.3 Adjustment to the Agreement provides a mechanism for the parties to request additional monies or time as a result of the subsequent protocol documents requiring additional services or efforts not initially contemplated by the contracting parties.

A chart provided in Section 3.1 identifies the highest level types of digital data anticipated on a project to include: "Project Agreements and Modifications,

Project communications, Architect's preconstruction submittals, Contract Documents, Contractor's submittals, Subcontractor's submittals, Modifications, Project Payment documents, Notices and claims, and Building Information Modeling." (While the chart includes "Building Information Modeling" as a distinct type of digital data, Article 4 is dedicated entirely and specifically to BIM, as is the associated protocol AIA – G202™ (2013) *Building Information Modeling Protocol Form*.) The chart then requires the parties to indicate the applicability, or non-applicability of each digital data type to the project. Subsequently, the parties must provide a "detailed description of the anticipated Digital Data identified."[185]

Following this preliminary identification of the types of digital data anticipated on the project, the parties are required to establish specific protocols related to "Authorized Uses" of digital data. The timeline for establishing those protocols is "As soon as practical after execution of the Agreement."[186] In contrast to the other major industry BIM form document ConsensusDocs 301 *BIM Addendum*, and several publically available guidelines as discussed above[187] which have a 30 day requirement for completing a project's BEP (or similarly named requirement), the AIA has chosen to leave the execution of the required protocol documents intentional, yet open, in order to meet different project's different needs, schedules, and circumstances.[188]

Article 3 further establishes that, unless otherwise noted, the architect will be responsible for preparing and distributing the protocol in the form of AIA – G201™ (2013) *Project Digital Data Protocol*.[189] This requirement echoes the primacy of the architect in the pre-cursor E202™ (2008). Each party also agrees to "memorialize their agreement in writing" to the protocol.[190] The parties further agree to review and, as needed, revise the protocol at appropriate project intervals in conjunction with other project participants.[191] The ability of the parties to revisit and revise the protocol, as well as memorialize their agreement to the protocol in writing, enables the protocol to remain a fluid, living document through the course of the project without requiring explicit amendment to the E203 exhibit and associated governing agreements.

Section 3.4 Unauthorized Use establishes that if a party receives digital data prior to the execution of the protocol, then that party cannot rely on the digital data. Should they do so, it is at their own risk and without liability to the other party. Similarly, after the establishment of the protocol the parties are bound to its terms, and use of the data in any manner other than prescribed in the protocol is done at the sole risk of the party using the data.[192]

Having enumerated a fair number of clauses for identifying the anticipated types of digital data on the project and then preparing the way for establishing specific protocols and approved uses of that information, Article 3 then poses a logical, significant question – do the parties intend to utilize a centralized electronic document management system to organize any or all of this data? Section 3.4 requires the parties to choose between two check boxes, one signifying their intention to use such a system, the other indicating they do not, with no choice by the parties defaulting to the latter.

While beyond the scope of this work to provide a complete analysis of centralized electronic document management systems, the topic is deserving of some commentary. At present, the amount of non-model digital data on even the most sophisticated of BIM/VDC-enabled projects far surpasses the amount of model data. Emails, reports, submittals, contract documents, payment documents, and so on – the list of non-model data is substantial and voluminous. Accordingly, the marketplace has, and continues to produce, any number of centralized electronic management systems seeking to bring order and efficiency to the vast mountains of non-model digital data produced during the course of designing and constructing things.

Some of these systems are pure *document* management systems seeking to bring order and accessibility to documents, but are either limited or unable to provide *data* management – the capacity to mine and leverage the specific content of those documents. Others are data management systems, able to provide deep insight to content, but that insight is specific or weighted towards a particular domain or document type – say emails or CAD drawings. Still others offer some combination of document and data management, but are focused on the professional needs of designers. Others are conversely geared towards the construction market. Further, construction-centric systems may focus on the macro-level needs of project management, but be unable to address the micro-level needs of superintendents charged with managing construction execution in the field.

Perhaps most significant for the purpose of discussion here, each of these distinct centralized electronic management systems will address BIM differently, if at all. Some offer deep integration between model and certain types of non-model data. Others offer no integration, but provide for basic organizational capabilities of models at the branch level. Yet others provide comprehensive organization and version control of models at the root level. Accordingly, given the significant effort to bring structure to digital data in Article 3, combined with the historical primacy of non-model documentation in the resolution of claims and disputes, parties would be well-served to consider the requirement of Section 3.4 in real time parallel with every single decision made across the exhibit and both associated protocols (BIM and digital data).

In much the same manner Article 3 was dedicated to establishing general purposes and preliminary intentions related to digital data, Article 4 is dedicated to structuring the same with regards to BIM. At the onset, the parties must choose between one of two, and only two, check boxes indicating how they intend to utilize BIM on the project.[193]

The first check box indicates that the parties intend to use BIM only to fulfill the obligations of their agreement and that there is zero expectation that the model will be utilized by any other project participants. Accordingly, the remaining sections of Article 4 have no effect on the parties. This could mean that if an owner and architect attach the E203™ to their agreement and select the first check box, then there is no obligation for the architect to further share his BIM content, even with his own consultants.

Alternatively, by selecting the second check box the parties indicate that they intend to utilize the model as per the terms set forth in the remaining sections of Article 4. Similar, though not identical, to the format of Article 3, those remaining clauses provide a structure for utilizing the model with additional project participants that is ultimately memorialized in the second protocol, G202™ (2013) *Building Information Modeling Protocol Form.*

Section 4.2 Anticipated Building Information Modeling Scope provides a blank dual-column fill-point for the parties to establish which portions of the project will be modeled and by whom. For example, this section could, at a high level, indicate those portions of the project to be modeled by particular design team firms, as well as those portions of the project to be modeled by particular construction team firms. As has been noted throughout, the E203™ is intended to be completed at the same time as governing agreements in an effort to best align team expectations with contractual realities. Section 4.2 thus provides an up-front opportunity for all stakeholders to begin to articulate where BIM starts, ends, and overlaps in the design–construct–operate continuum. Section 4.3 and Section 4.4 subsequently allow parties to indicate preliminary expectations regarding authorized uses of the model and any ancillary modeling activities (e.g. performance simulations), respectively. Authorized uses will be ultimately memorialized in the accompanying protocol, and follow generally the same themes (e.g. analysis, cost estimating, schedule, coordination, etc.) as provided for in the forerunner, the E202™.

Mirroring the requirement for completing the digital data protocol, the parties are required to complete the G202™ (2013) *Building Information Modeling Protocol Form* "As soon as practical following the execution of the Agreement."[194] Similar to the precedent initiated in the E202™ (2008), Section 4.5.1 requires that the protocol addresses topics such as identification of MEAs and definitions for LOD. In alignment with other portions of the E203™ geared towards clarifying post-construction expectations, Section 4.5.1.8 requires the project participants to address "Anticipated Authorized Uses for facilities management or otherwise, following completion of the Project."

By default, the project participants agree that the architect shall prepare and distribute the BIM protocol form, unless that responsibility is assigned to another project participant.[195] Likewise, the parties must memorialize in writing their agreement to the latest BIM protocol form and agree to revisit and revise the form as needed.[196] Being a BIM-specific document, the parties also agree to incorporate the AIA G202™ within the model itself, or otherwise attach it.[197] Rules governing unauthorized use of the model prior to, and after execution of, the AIA G202™ mirror those regarding digital data and execution of the G201™.

Section 4.8 Model Management incorporates many of the model management responsibilities and requirements historically enumerated in Section 2.4 of the predecessor, AIA E202™ (2008). For example, in the revised AIA E203™ (2013) the default model manager for all project phases is still the architect, and requirements are similarly broken out into initial, macro-level responsibilities such as determining the model origin point and coordinate system,[198] as well as on-going responsibilities such as collecting and aggregating incoming models and managing access rights.[199] Responsibilities for model archives are likewise similar to the AIA E202™ (2008), although the revised AIA E203™ (2013) removes the explicit requirement of the former to archive two sets of models – one being a collection of individual models as received from the MEA, the other being an aggregate of those models.

Given the several revisions and additions contained in the AIA E203™ dealing specifically with post-construction BIM expectations, it seems only fitting that Article 4 concludes with Section 4.9 Post-Construction Model. While not voluminous, Section 4.9 arguably provides one of the most pertinent improvements to the AIA E202™ (2008) by providing an arena for the Parties to clearly articulate BIM expectations, and by implication the more important issue of compensation, regarding facilities management. Section 4.9 dictates that "The services associated with providing a Model for post construction shall only be required if specifically designated in the table below as a Party's responsibility." The given table then enumerates several typical post-construction uses of a model including: remodeling, wayfinding and mapping, asset/furniture, fixtures, equipment (FF & E) management, energy management, space management, and maintenance management. The parties must then identify if the given use (or any other use added to the table by the parties) is applicable to the project, who is responsible, and subsequently provide a detailed description of the associated requirements and services for those deemed applicable.[200]

As with the parties' potential decision regarding the use of a centralized electronic data management system for digital data, the requirements and services for delivery of post-construction models can be complex. Deliverable requirements for a project have the potential to vary greatly depending upon any number of factors, including the given owner's level of BIM knowledge, their existing internal systems, or their organizational capacity. For example, a sophisticated owner may have a number of internal facilities and operations systems for separately managing: (1) preventative maintenance on mechanical equipment, (2) square footage for departmental business unit charge-backs, and (3) fixed/movable assets and their associated depreciation for tax and accountancy purposes. Each of these three example systems may wish to retrieve certain inputs from model content, or data located in any centralized electronic data management system identified in Section 3.5.1, or both.

In some instances, the coordinated D-BIM may be the best source of information. In others, the as-built C-BIM is perhaps best suited. And in still certain others, perhaps a conformed design BIM provides the clearest path to graphical data or 3D visualization for system inputs. Likewise, non-model digital data housed in the data management system might bring the most efficiency and cost savings to the owner post-construction. Alternatively, a less robust owner may manage their real estate portfolio more simplistically, utilizing a basic spreadsheet and the "mental hard drives" and experience of seasoned personnel. This is all to say that the structure required for a successful BIM to facilities management transition requires significant planning and execution.

With the 5-year revision and re-configuration cycle of its predecessor, the E203™ (2013) appears to be meeting aggressive market demands for BIM guidelines and best practices. The essentially wholesale retainage of the requirements and duties from the predecessor E202™ (2008), including LOD and authorized use assignments, duty to inform regarding conflicts, and so on, suggests a standard of care. The addition of new features and potential requirements, including the tri-document arrangement of exhibit and two protocols, consideration of digital non-model information in parallel with BIM, potential uses of BIM for facilities, and so on, suggest that the standard of care has begun transition to its next phase.

3.4.3 AIA – G201™ (2013) Project Digital Data Protocol

The AIA G201™ (2013) *Project Digital Data Protocol* is the first of two protocols for use in conjunction with the E203™ (2013) exhibit. As the title suggests, the G201™ (2013) is a tactical document wherein the project participants establish more specificity around procedures and protocols regarding digital data. Given the high level integration between the G201™ (2013) and the E203™ (2013) the AIA has provided a marginal note on the title page indicating this relationship and reminding parties that the latter will be incorporated into respective governing agreements. Accordingly, the title block references the given project and not a specific governing agreement. While the protocol is intended to be controlling on project participants, it is also envisioned as flexible enough to meet any changes or on-going project demands. The title block reflects this vision with, "Protocol Version Number" as a fixed sub-header. And, as discussed above in the review of the E203™ (2013), Section 1.3 Adjustments to the Agreement of that document establishes procedures for requesting and negotiating possible changes to compensation and time that parties may feel entitled to as per requirements of the most current protocol version.

Structurally, the G201™ (2013) is comprised of three articles:

Article 1 General Provisions Regarding use of Digital Data

Article 2 Digital Data Management Protocols

Article 3 Transmission and Use of Digital Data.

Formalizing the integration between the G201™ and the E203™, Article 1 requires a listing of each project participant that has incorporated the project-specific E203™ exhibit into their agreement for the project. Recall that the E203™ is intended to serve as the initial global source document for establishing macro-level intentions regarding digital data and BIM for all project stakeholders. In addition to listing each project participant that has incorporated the exhibit, Article 1 further requires a listing of the names and contact details for each individual(s) responsible for implementing the given digital data protocol at each project participant. Article 1 concludes with recognition that the terms found in the G201™ shall have the same meaning as set forth in the E203™.

Article 2 begins by requiring project participants to detail the requirements for the centralized electronic data management system, if such a system was indicated in Section 3.5.1 of the E203™ (2013). Given the variety and complexity of data management systems previously noted, the possibility exists that a project might require more than one system, perhaps concurrently or relative to a particular phase of the project. While written in the singular, it seems fair to assume that the parties would use Article 2 to detail the specific requirements of any and all data management systems to be utilized.[201]

Having specified the system(s) requirements, project participants are next required to detail if any start-up training, or on-going training or support is necessary to implement and maintain the system.[202] While not requiring any such training or support, given the cultural phenomena in the US AEC industry for projects to often either not fund, or under-fund, training or maintenance associated with technology in general, Article 2 nonetheless forces project participants to at least consider any implications of foregoing training and support.

Article 2 concludes by requiring project participants to articulate requirements, if any, for the central electronic data management system(s) to store digital data during the course of the project, as well as preserve and archive digital data during the course of the project and following final completion.[203] Given the importance of documentation and correspondence in claims and dispute resolution, project participants should carefully consider storage and archiving. For example, a licensing agreement for a given data management system may provide access to project participant B through a governing system license held by project participant A. In such a case, ownership and access to data might be, barring intervention by others, controlled by A at the conclusion of a project.

With controlling digital data management protocols now established, Article 3 of the G201™ (2013) articulates the transmission methods and authorized uses of digital data. Echoing the matrix structure of the model element table seen in the predecessor E202™ (2008), Section 3.1 Digital Data Protocol Table offers the project a visual aid for quickly determining how various types of digital data (e.g. meeting notices) are to be transmitted (e.g. via email) and how they may be used (e.g. stored and viewed only).

3.4.4 AIA – G202™ (2013) Project Building
Information Modeling Protocol Form

The AIA G202™ (2013) *Project Building Information Modeling Protocol Form* is the second of two protocols for use in conjunction with the E203™ (2013) exhibit. For professionals familiar with implementing the predecessor E202™ (2008), the G202™ will be familiar insofar as the document incorporates the more "well-known" features of the former, namely LOD definitions, and the model element table.

Structurally, the G202™ (2013) is organized into three articles:

Article 1 General Provisions

Article 2 Level of Development

Article 3 Model Elements.

Echoing a format similar to the G201™, Article 1 requires identification of the project participants that have incorporated the project specific E203™ (2013) exhibit into their agreements. Further, the specific individual(s) who will be responsible for implementing the protocol must be identified and their contact details provided.[204]

Section 1.2 next requires project participants to identify what "information and other data sets" comprise the model. Distinct from identification of specific model elements and their MEAs, this section allows the project participants to determine what other non-model information collectively forms the model. Examples might include: energy analysis models, cost-estimating databases, operations and maintenance manuals, and so on.

Project participants must next identify which, if any, collaboration protocols for utilizing the model will be used. Examples of potential items to address include: "communications protocols, a collaboration meeting schedule, and any co-location requirements."[205] Given that most typical design and construction contracts include provisions regarding collaboration and meeting requirements, Section 1.3 requires project participants to identify any BIM-specific requirements. For example, this might require the architect to establish a D-BIM coordination meeting schedule that is separate from any typical design coordination meeting requirements in his contracts with an owner and/or design consultants. This is not to say that BIM might not also be utilized as part of those typical coordination meetings; however, a dedicated schedule for BIM matters, for example resolving design-side clashes, might now be required.

It is also possible that a collaboration protocol could include a collaboration technology – hardware, or software, or both – to enhance effectiveness. Project participants will be required to articulate any specific requirements around any such solutions. Likewise, as was previously discussed, Section 3.5.1 of the E203™ (2013) requires project participants to determine if a centralized electronic document management system will be utilized on the project. Project participants will likely avoid confusion by clearly mapping any known overlap(s) or gap(s) between these systems.

Any hardware or software requirements associated with a possible collaboration technology would be identified in Section 1.4 Technical Requirements, which requires project participants to identify all hardware and software requirements "relating to the utilization of Building Information Modeling" on the project. And, like the G201™, the G202™ requires project participants to identify any training or support mechanisms needed to establish and maintain collaboration and technical requirements.

Similar to the predecessor E202™ (2008), the G202™ allows project participants to indicate what model standard, if any, will be utilized. The re-cast G202™ has, however, removed mention of the *National Building Information Modeling Standard* as a possible model standard.

Section 1.7 Model Management Protocols and Processes provides a chart in which project participants: (1) identify which protocols and processes are applicable to the project, and (2) where detailed information relative to each protocol or process can be found. By default those detailed descriptions are located in Section 1.8, or in an attached exhibit to the G202™. Reinforcing the separate, yet supporting structure of the E203™ (macro-level BIM intentions) and G202™ (detailed BIM protocols), the specific requirements enumerated in Section 1.7 (e.g. model origin point, naming conventions, etc.) mirror those found in Section 4.8.2 Model Management Protocol Establishment of the E203™. Along the same lines of reinforcement, Article 1 concludes by stating that the terms utilized in the G202™ have the same meaning as set forth in the E203™.

For professionals already familiar with the E202™, Article 2 of the G202™ will be immediately recognizable in that the definitions for each progressive LOD (100 to 500) are incorporated essentially wholesale from the former. Any differences from the inaugural E202™ definitions are refinements in language and present no substantive changes. For example, the definition for LOD 300 conveys the same minimum content requirements, and "specific" remains the operative word, "The Model Element is graphically represented within the Model as a specific system, object or assembly in terms of quantity, size, shape, location and orientation. Non-graphic information may also be attached to Model Element."[206] Similarly, LOD 500 conveys essentially the same requirements regarding as-built Model Elements, but uses new clarifying language, "The Model Element is a *field verified* representation in terms of size, shape, location, quantity, and orientation. Non-graphic information may also be attached to the Model Elements" (emphasis added).[207]

As with the LOD definitions, the G202™ likewise incorporates (essentially unaltered) the associated authorized uses from the E202™. Each authorized use dictates a project participant's use of, and reliance on, model content associated with each LOD. The G202™ has, however, added a new authorized use of Coordination for LODs 200, 300, and 400. The coordination definition is specific to each said LOD and increases in specificity. For example, LOD 200 states "The Model Element may be used for coordination with other Model Elements in terms of its size, location, and clearance to other Model Elements." LOD 400 authorizes a model element to be coordinated to the same terms, with the addition of "fabrication, installation and detailed operation issues."[208] The addition of an authorized use regarding coordination provides, amongst other possible benefits, a structure for intelligently incorporating long-term facilities and maintenance use requirements into a project at appropriate intervals.

For LOD 500, the general usage definition used in the E202™ related to using content for maintaining, altering, and adding to the project as granted by any applicable license(s) has been removed in the G202™. This was likely done to coordinate and reinforce Section 2.3 of the E203™ which expanded a party's usage rights beyond design and construction to include "using, maintaining, altering and adding to the Project." As previously noted, that language matches Section 7.3 of the AIA's B101™ *Owner Architect Agreement (2007)* that provides owners with a nonexclusive license to use an architect's instruments of service.

The G202™ concludes by: defining reliance, establishing duty with regard to coordination and model refinement, and incorporating the model element table from the E202™.[209] In a slight, but pointed change from the E202™, the definition for reliance is built around project milestones as opposed to project phases.[210] As noted above, this was most likely done in response to acknowledging the fact that the LOD progression is not necessarily intended to be perfectly analogous to traditional project phasing. That is, LOD 100 does not necessarily equal SD, LOD 200 does not necessarily equal DD, and so on. Early implementation of the LOD scheme in US practice often led projects to attempt to equate LOD with traditional project phasing in a wholesale fashion, that is, at the conclusion of DD an entire model should be, say, LOD 200. The industry was quickly confronted with the fact that a model at any given point in time can include model elements across systems at various LOD's. The fluid nature of design and construction may result in certain systems being "ahead" or "behind" other systems from a strictly LOD perspective. Thus, a model would unlikely be, in its entirety, a LOD 200 model. Accordingly, Section 3.1.1 establishes that project participants can only rely on model element accuracy and completeness as consistent with the LOD designation in the model element table for the stated milestone, even if the model element exceeds the required LOD.

With reliance defined, Section 3.1.2 then establishes a duty with respect to identification of conflict stating, in part:

> Where conflicts are found in the model, regardless of the phase of the Project, or LOD, the Project Participant that identifies the conflict shall promptly notify the Model Element Authors and the Project Participant . . . responsible for Model Management. Upon such notification, the Model Element Author(s) shall act promptly to evaluate, mitigate and resolve the conflict in accordance with the processes established in Sec. 1.7.7, if applicable.

The reference to Section 1.7.7 is the line-item model management protocol for "Design coordination and clash detection procedures." If the phrase "clash detection procedures" is to be interpreted generally across all project participants, phases and milestones, and not specific to design coordination, this now technologically-driven duty to reveal conflict would appear to relate – at least abstractly – to the established construction law concept of a contractor's "duty to inquire."[211]

As with the E202™, the model element table in the G202™ provides a visual control matrix for project participants to establish reliance and coordinate modeling work across disciplines.[212] Functionally, model elements are assigned a

MEA, and associated LOD(s) across project milestones. With the G202™, the definition for MEA (as defined in the E203™) has been expanded, "The Model Element Author is the entity (or individual) responsible for managing and coordinating the development of a specific Model Element to the LOD required for an identified Project milestone, regardless of who is responsible for providing the content in the Model Element." This expanded definition directly addresses the notion of *responsible control* with respect to BIM workflows, in that the MEA assignment does not change who is in responsible control of the design.[213]

Whereas the E202™ defaulted to the CSI Uniformat™ classification system for organizing model elements, the G202™ sample document leaves determination of the classification system to project participants. And, as noted above, with project milestones, not phases, controlling reliance at given points in time, project participants are required to identify said milestones in completing the model element table.

3.4.5 *ConsensusDocs – 301*™ Building Information Modeling (BIM) Addendum

ConsensusDocs™ is a consortium of more than forty associations across the entire spectrum of the design–construct–operate industry with a catalog of more than 100 standard form contract documents.[214] ConsensusDocs™ was the first industry association to offer a standard BIM contract document, issuing the 301 *Building Information Modeling (BIM) Addendum* in 2008 just slightly ahead of the AIA's E202™ *Building Information Modeling Protocol (2008)*. ConsensusDocs™ maintains a 5-year, or sooner, cycle for the review and, where necessary, update to any document.[215]

The ConsensusDocs™ 301 (2008) is comprised of six sections (see Appendix B):

Section 1 General Principles

Section 2 Definitions

Section 3 Information Management

Section 4 BIM Execution Plan

Section 5 Risk Allocation

Section 6 Intellectual Property Rights in Models.

As several noted construction law practitioners (Lowe & Muncey, O'Conner & Hurtado, Larson & Golden, etc.) have described, early BIM proponents were often confronted by objections that BIM tools and processes could not be effectively used on projects utilizing typical design–bid–build contracts. Emerging technologies were portrayed as too collaborative, too porous, and too unpredictable to support the clearly separated inter-, and intra-distinctions of the design and construction domains. These objections were, quite logically, rooted in long-developed construction law concepts including, for example, privity of contract, responsible charge of design, and *Spearin* doctrine warranty.

Thus, prior to wading directly into the expected technological weeds and details concerning the use of BIM/VDC tools and processes, the general principles contained in Section 1 specifically address typical legal objections. For example, Paragraph 1.1 clarifies that the 301 (2008) "does not effectuate or require a restructuring of contractual relationships or shifting of risks between or amongst the Project Participants." Likewise, Section 1 states that the *Addendum* does *not*: create privity of contract in and of itself,[216] relieve the architect (or engineer) of their obligation as the person in responsible charge of the design,[217] nor diminish an owner's warranty to any party regarding the adequacy and/or sufficiency of the design.[218]

As the document title indicates, the 301 (2008) is intended as an addendum and Paragraph 1.3 thus establishes requirements for appending and incorporating the *Addendum* into both governing contracts and affiliated contracts, with each type of contract in turn containing flow-down provisions to sub-consultants and subcontractors. A governing contract is defined as "the agreement to which this *Addendum* is attached and in which it is incorporated, but excludes an Affiliated Contract."[219] An affiliated contract is defined as "any contract relating to the Project to which an identical Addendum is attached and . . . incorporated, other than the Governing Contract."[220] Section 1 also contains a general precedence clause stating, "In the event of an inconsistency between this Addendum and the Governing Contract, this Addendum shall take precedence."

Prior to analyzing the remaining clauses in Section 1, it is worthwhile to present the multiple definitions for various types and states of models within the 301 (2008), and how such models relate to typical definitions for drawings, CDs, and contract documents.

In alignment with general industry vocabulary, the 301 (2008) defines a model as "a three-dimensional representation in electronic format of building elements representing solid objects with true-to-scale spatial relationships and dimensions. A Model may include additional information or data."[221] The *Addendum* also defines a contribution as the "expression, design, data or information" that a project participant creates and shares with other project participants for use with a model.[222] Thus a contribution could be a model or 2D drawings not derived from a model.

Recognizing that both designers and contractors might produce models for a given project, definitions for both design model and construction model are presented. A design model is defined as

> a Model of those aspects of the Project that (a) are to be modeled as specified in the BIM Execution Plan . . . and (b) have reached the stage of completion that would customarily be expressed by an Architect/Engineer in two-dimensional Construction Documents. This shall not include Models

such as analytical evaluation, preliminary designs, studies, or renderings. A Model prepared by an Architect/Engineer that has not reached the stage of completion specified in this definition is referred to as a Model.[223]

Before continuing it is noted that the 301 (2008) does not define CDs. Accordingly, parties might turn to the governing contracts for a definition. By way of general example, the AIA B101™ 2007 *Standard Form of Agreement Between Owner and Architect* indicates that CDs consist of, in part, drawings and specifications setting forth detailed "levels of materials and systems and other requirements for the construction of the Work."[224] As the AIA notes, CDs need to contain sufficient detail to obtain competitive bids and to communicate to contractors the actual result to be achieved.[225] Significantly, the B101™ 2007, as the revision/compilation of two earlier AIA documents, also added a new element to the definition of CDs: "The Owner and Architect acknowledge that in order to construct the Work the Contractor will provide additional information, including Shop Drawings, Product Data, Samples and other similar submittals, which the Architect shall review."[226]

As others have noted, while the AIA has always addressed shop drawings, 2007 represents the first time those items have been discussed in conjunction with CDs. It has been suggested that the likely impetus for this inclusion is to address the fact that shop drawings are a necessary pre-requisite to starting construction, but also the "trend in the construction industry toward architects detailing Construction Documents less fully than in past practice."[227]

In light of typical contractual relationships between an architect and his sub-design consultant(s), the *Addendum* has created a full design model. A full design model is a model that contains coordinated design models from structural, MEP, and other design models designated to be produced by the design team.[228] While the *Addendum* is silent on a definition for "coordinated" it strikes one as reasonable to assume that term to have the customary meaning: that MEP systems (including any specialty systems) and architectural systems and structural systems have been carefully reviewed by those in responsible charge to make certain that elements from any system or discipline do not spatially interfere with each other.[229]

With design model and full design model so defined, the 301 (2008) also designates a design model, by default, to be a contract document. Paragraph 2.3 states, "Contract Document, as defined in the Governing Contract, is modified to include all Design Models, unless otherwise specified in the BIM Execution Plan." Additional paragraphs in the *Addendum* further clarify and expand upon this point,[230] but the *Addendum* is clear: by default a design model is a contract document.

Since design models as defined have parity with typical CDs, a construction model is, therefore, defined as, "a Model that . . . utilizes data imported from a Design Model or, if none, from a designer's Construction Documents; and . . . contains the equivalent of shop drawings and other information useful for construction."[231]

With constituent model types separated along typical design/construct boundaries, the 301 (2008) provides for two levels of amalgamation of models across those boundaries. At the first level of coordination is a federated model. As the name implies, a federated model means a model of linked, but distinct, models that do not lose their "identity or integrity by being so linked."[232] The federated model thus becomes a technical realization of the clear distinction between traditional architect and contractor responsibilities and duties maintained in Section 1. The linking of models does not blur the line separating design and construction at any point during design and construction. Responsible control of the design is maintained by the architect. The contractor and subcontractors maintain a duty to perform in a workman-like manner, but do not perform design services.

At the second level of amalgamation is a project model which is established by the linking of a full design model and one or more construction models. A project model thus has potential for project requirements including as-built conditions as will be discussed in due course below.

In addition to the preceding definitions concerning models, the 301 (2008) provides a two part definition concerning drawings:

> Drawings means (a) those two-dimensional plans, sketches or other drawings that are Contract Documents under the Governing Contract and are created separately from, and are not derived from, a Model and (b) those two-dimensional projections derived from a Model supplemented with independent graphics and annotations specified by the Parties to be Contract Documents.[233]

Thus informed on defined terms the remaining model-specific elements of Section 1 are clarified. Paragraph 1.8 establishes the precedence of a design model in the event of a conflict with any other model. Additionally, by default, a design model is understood to not provide "the level of detail needed in order to extract precise material or object quantities."[234] Similarly, "the dimensional tolerances provided by the Contract Documents in the Governing Contract shall apply to dimensions in the Model."[235]

Section 1 also establishes a project participant's duty to inform with respect to discrepancies between models or between a model and other contract documents. "If any Project Participant becomes aware of a discrepancy between a Model and either another Model or another Contract Document, such Project Participant shall promptly notify the other Party or Parties to that Project Participant's Governing Contract and the Information Manager."[236]

As already seen, the *Addendum* maintains the traditional obligation of the architect as the person in responsible charge of the design, and the owner's customary *Spearin* warranty. Therefore, this duty to notify aligns with traditional notions of reporting errors, and places no additional duty on a contractor to otherwise determine the plans and specifications to be buildable.

Section 2 defines all relevant terms for the 301 (2008). Rather than present here any terms that haven't already been discussed, those remaining definitions will be examined in context below. Towards that end, Section 3 is dedicated to information management, where information management is defined as the protocols related to information and information systems relative to BIM, that is, the technical software and/or solution for managing the various models on the project.[237] For loose comparative purposes, Section 1.7 Model Management Protocols and Processes found within the AIA G202™ (2013) *Project Building Information Modeling Protocol Form* establishes similar project requirements.

In contrast to the G202™ (2013), however, the 301 (2008) does not appoint the architect as the default administrator of such a system. The information manager, or "IM" as this entity is referred to in the 301 (2008), is appointed by the owner, or its designated representative. The owner may appoint one or more IMs, including the architect, the contractor/construction manager, and/or another entity. By default, all compensation and related costs for the IM are paid by the owner.[238]

The responsibilities of the IM include providing or procuring an information management system capable of delivering functionality such as: creating and maintaining user accounts,[239] assigning access rights,[240] and applying access controls "so that only authorized users of the Model can access only the data they are authorized to access."[241] The information management system is also required, at a minimum, to record a data/time entry for each data entry by users of the system, including downloading of Models.[242] This requirement is significant given the importance of project documentation and re-constructed timelines in resolving disputes. The itemized transparency of Model transactions offered by a robust information management system is a useful tool in investigating any potential claim either centered on, or clarified by, BIM.

Section 4 BIM Execution Plan is arguably the best known portion of the 301 (2008). Regardless of the term's provenance as industry vocabulary, "BIM Execution Plan" has become both a familiar and expected component of any project competently utilizing BIM/VDC. A BEP is just that – an execution plan that establishes the roles, responsibilities, protocols, and specific details required of project stakeholders with regard to BIM.

In terms of contract administration, the *Addendum* requires that, "As soon as practicable, but in no event later than thirty (30) days after the latter of the execution of the Contract between" the owner/architect, or the owner/contractor, all project participants meet and develop a BEP.[243] Once agreed upon, the BEP and any modifications become an amendment to the 301 *Addendum* itself, which is incorporated into the governing contracts as per Paragraph 1.3. In ideal practice, this process enables the parties to incorporate their general BIM requirements via the *Addendum* into governing contracts at the time of their execution, and then provide a window of development time for the project participants to develop details required by the BEP. Again in loose comparison to the AIA BIM documents, the re-constituting of the single document E202™ (2008) into the triumvirate of the E203™ (2013), and G201™ (2013) and G202™ (2013) protocols in its revision is intended to provide similar flexibility: enable owners, architects, and contractors to set macro-level BIM expectations and intentions at the execution of governing contracts, with micro-level BIM protocol details being established in protocols at more appropriate intervals in the project timeline.

In the 301 (2008), establishing those micro-level details in the BEP begins with "Identification of what Models are to be created, the purpose(s) each Model is intended to serve, and which Project Participant(s) is (are) responsible for creating each model."[244] The overall intent of the BEP requires the project participants to establish "the content of each Model and the required level of detail at various Project milestones."[245] Required details for model content include: geometric and spatial data, object properties, object constitution data, and provisions for object parameters concerning cost and schedule data.[246] As one would expect, the BEP also enumerates an extensive list of technical requirements including: establishing a common coordinate system,[247] unit conventions,[248] file and object naming conventions,[249] and so on.

In addition to requiring details specific to models, the BEP also requires project participants to further clarify the status of various models or derivatives from those models as contract documents. For example, Paragraph 4.3.3 requires the parties to identify which design model or models are not contract documents. Recalling that, by default, Paragraph 2.3 modifies the definition of contract documents in any governing contract to include all design models, Paragraph 4.3.3 thus provides for any needed exclusions.

Likewise, the BEP establishes protocols for submission and approval of models, including any "electronic stamping," and the process for designating a model as a design model.[250] Since a design model is defined, in part, as a model that has reached the level of completion customarily expressed by an architect in 2D CDs, and the *Addendum* does not offer a definition for CDs, this requirement provides project participants the opportunity to qualify exactly what customary CDs means for the given project.

The BEP also requires project participants to determine what model(s) shall be part of the record documents for the project.[251] Likewise any procedures for "confirmation of field changes through an as-built Project Model" must be established Paragraph 4.3.26.

If one concedes Section 4 BIM Execution Plan as the most popularly recognized portion of the *Addendum* in general, then Paragraph 4.3.11 Contributor's Dimensional Accuracy Representation is perhaps the specific clause most discussed – and debated – on projects utilizing the 301 (2008). Here, project participants must choose one, and only one, of four available choices related to the dimensional accuracy of BIM content. Dimensional accuracy is, at the end of the day, the fundamental question on any BIM/VDC project. If forced to rely, on what does one rely – drawings or models? Each of the three choices in Paragraph 4.3.11 progressively limits the dimensional reliability of BIM, with the fourth "other" choice leaving the door open (but likely rarely entered) for project stakeholders to draft their own language for BIM reliability.

Prior to making their single selection, any choice is qualified as: being limited to other parties to the governing contract, in accordance with the standard of care applicable to the contributor, and effective at the time the model has parity with traditional 2D CDs.[252]

The first check box is the most progressive in terms of BIM/VDC, stating that a contributor "represents that the dimensions in its Contribution to a Model are accurate and take precedence over the dimensions called out in the Drawings or inferred from the Drawings. Details and components that are not represented in . . . a Model must be retrieved from the Drawings." The second choice narrows BIM reliance, dictating that each contributor "represents that the dimensions in its Contribution to a Model are accurate to the extent that the BIM Execution Plan specifies dimensions to be accurate, and all other dimensions must be retrieved from the drawings." The third choice is the least BIM/VDC progressive stating, "Contributors make no representation with respect to the dimensional accuracy of the Contributor's Contribution to a Model. A model can be used for reference only and all dimensions must be retrieved from the Drawings." To the dismay of many early BIM/VDC proponents, the third option was often the choice typically selected by project participants in the early ascent of the adoption curve. The typical deference to drawings over models deserves increased attention within the industry.

While not to suggest the impossibility of project/situational variance, current BIM authoring software solutions overwhelmingly deliver direct parent–child relationships between models and the 2D drawings extracted from those models.[253] The dimensions of the model *are* the dimensions of the drawing. Achieving a different result would very likely require an active choice on the part of the end user to break the established relationship. Even in situations where a particular authoring tool does not offer bi-directional associativity between the 3D model and associated 2D drawing, competent workflows are typically built around beginning with the model to then generate static drawings. When changes are required in such non-bi-directional situations, those changes are made to the model first, with the model then being "refreshed" to re-create the 2D drawing. Similarly, where any issues might exist in the purely visual presentation of extracted or projected 2D drawings from a model, they are best categorized as minor graphic infractions as opposed to substantive dimensional variances. For example, two pipes in an extracted plan view showing atop one another as opposed to offset to reveal the elevation of each, would still, in other parametrically derived sectional views from the same model – not to mention holistically in a 3D view – appear at their correct elevation heights.

Lastly, as already noted, even in situations where details are not part of models, competent best practice utilizes the underlying framework of the model as the armature upon which additional 2D graphic symbology is applied. For example, envision a typical wall section being parametrically cut through a model where the sectional cut line passes through a window in the wall. In detailing the window portion of the resultant wall section (e.g. showing waterproofing membranes, step-flashing, etc. as required by scale) the end user can usually: directly apply 2D detail line work within the native BIM authoring tool that is associated – and where needed – "locked" to the parametric wall cut at the location of the window, or link and lock a static 2D detail line drawing from outside the model (e.g. from a legacy library of standard non-BIM details), or some utilize some combination thereof. The *in situ* context and (where applicable) "locked" connection between the underlying model armature and any applied 2D detail line work of either/any scenario inherently delivers better coordination and dimensional accuracy than a typical alternative which relies upon the skill of the draftsperson to coordinate thousands, if not millions, of static lines across hundreds, if not thousands of drawings.

By way of enlarged context beyond the 301 (2008), several publically available BIM standards have expressly commented on the relationship between models and 2D drawing projections/extractions. The Indiana University *BIM Guideline* requires that "All drawings, schedules, simulations, and services required for assessment, review, bidding and construction shall be extractions from [a] model."[254] The *New York City Department of Design and Construction BIM Guidelines* similarly requires that plans, sections, elevations, details and schedules be created from a model.[255] Likewise, the State of Wisconsin BIM *Guidelines* state, in relevant part:

> Architects and Structural Engineers shall use BIM Authoring software (see section 1.4). Building information models shall be created that include all geometry, physical characteristics and product data needed to describe the design and construction work. All drawings and schedules required for assessment, review, bidding and construction shall be extractions from this model.[256]

The BEP also requires project participants to establish protocols for the utilization of BIM for the RFI[257] and change order[258] processes. With RFI and change order documentation playing a critical role in dispute and claim resolution, this requirement couched with both, the static and time-based visual scheduling capabilities of BIM, and the information management system requirements of Section 3 relative to cataloging model transactions between and amongst all project participants, could prove extremely effective in evaluating disputes or claims.

Section 5 addresses risk allocation and includes clauses pertaining to issues, including: waiver of consequential damages, standard of care, duty to inform, insurance, and software defects. As a baseline, a party is responsible for any contribution(s) it makes to a model(s), or issue(s) that arises from access to a model. This responsibility flows down to project participants of a lower tier than the parties who are in contractual privity.[259] Accordingly, a subcontractor in an affiliated contract with the contractor remains responsible for contributions made to a model.

As noted by Lowe and Muncey, the waiver of consequential damages between the parties relating to models was of vital importance in the drafting of the 301 (2008). Without the mutual waiver of consequential damages – including but not limited to: rental expenses, loss of income of profit, principal office overhead and expenses, and so on – the parties may have foregone the use of BIM entirely as a result of increased and/or unknown risks.[260] Thus, the *Addendum* defers to the governing contract on the issue of waiver of consequential damages arising from a contribution, and requires each party to "waive claims against the other Parties to the Governing Contract for consequential damages arising out of or relating to the use of or access to a Model."[261] Similarly, parties involved in creating a model are not responsible for costs, expenses, liabilities, or damages resulting from the use of the model beyond the terms of the *Addendum*. In broad summary, if you add it to the model you own it. In return your contribution is protected by the explicit terms of use spelled out in the *Addendum*.

Paragraph 5.3 further clarifies the relationship between design models, contract documents, and the parties' representation of dimensional accuracy with respect to models and drawings as chosen in Paragraph 4.3.11. While project participants may rely upon the accuracy of information contained in any design models designated as a contract document, the dimensional accuracy of any contribution or model is still controlled by the selection in Paragraph 4.3.11.

Article 5 does not change the standard of care applicable to each party regarding the party's contribution to or use of a model. The applicable standard of care is determined by a party's governing contract or common law, as applicable.[262]

As discussed above, each project participant has a duty to inform with respect to the discovery or knowledge of inconsistencies between a model and either another model or contract document. Similarly, Paragraph 5.5 places a duty on each party to, amongst other efforts, minimize risk of claims and liability, "promptly report . . . any errors, inconsistencies, or omissions it discovers in its Model or the Project Model." Paragraph 5.7 speaks to insurance minimums. Unless otherwise agreed, the parties agree to obtain and maintain valuable papers and records coverage.

Perhaps informed and influenced by *MA Mortenson Co.* v. *Timberline Software Corporation*[263] in which a contractor's bid was significantly low as a result of an error in an estimating/bid preparation software package, Section 5 concludes by addressing software defects. In the event of a software defect impacting the creation, modification, federation or other use of a model or project model, the affected party is entitled to a time extension or other excuse from performance. However, any excuse is contingent up the party having avoided the delay or loss through the exercise of reasonable care in the first instance.[264]

Section 6 Intellectual Property Rights in Models is the final section of the 301 (2008). From a copyright perspective, the *Addendum* requires that each party warrant to other parties to the governing contract that it either holds the copyright to all of its contributions, or that it is licensed or otherwise authorized to use the contribution. Furthermore, each party agrees to indemnify and hold harmless the other in the event of third-party claims "arising out of or relating to, claims or demands relating to infringement or alleged infringement."[265]

Likewise, Paragraph 6.1 states that "each Party grants to the other Party or Parties to the Governing Contract (a) a limited, non-exclusive license to reproduce, distribute, display and otherwise use that Party's Contributions" for the purposes of the project. Sublicensing language also allows for similar use of contributions by project participants in affiliated contracts. The net-effect is to allow all project participants to governing and affiliated contracts to utilize models and contributions in the delivery of their contractual obligations on the project.

While the preceding provisions clarify project participants' use of contributions during the course of designing, constructing and otherwise delivering the project, the owner's use of the full design model post-construction is controlled by the governing contract between the owner and architect.[266] Given the interest of owners to utilize BIM deliverables for facilities management, many governing contracts, BIM addendums, and BIM guidelines are now being drafted accordingly.[267] As pertains to the owner's license generally, it can be terminated for failure to meet its project payment obligations if so adjudged by a court of law or arbitration.

3.5 Summary

The goal of this chapter was to provide supporting evidence for accurate expectations of the current standard of care and workmanlike performance with respect to BIM/VDC. This process began with an overview of these topics generally through samples of relevant language from current industry form contracts and example cases (Section 3.1). Next, BIM/VDC-specific source documents, drawn from across three "bands" of industry knowledge were reviewed and summarized. General foundational research included CIFE academic research, and the AGC *Contractors' Guide to BIM* (Section 3.2). This was followed by a review of publically available first generation BIM guidelines and standards including the State of Ohio *Building Information Modeling Protocol* and the *New York City Department of Design and Construction BIM Guidelines* (Section 3.3). Lastly, recent and current BIM/VDC-specific form contract documents issued by the AIA (e.g. E203 (2013)) and the ConsensusDocs 301 *Building Information Modeling Addendum* were examined (Section 3.4).

Collectively and individually these sources would appear to support expectations of standard of care and workmanlike performance regarding BIM/VDC that include:

- the use of BIM authoring tools as the primary production tool for producing more consistent and better coordinated 2D contract document sets;
- the use of Project Review tools to improve systems coordination via 3D clash detection.

Indeed, as previously noted, at one point during the course of writing of this book (June 2014), the National Council of Architectural Registration Boards (NCARB) posted commentary related to proposed changes to their Intern Development Program for training and preparing architects for licensure.

Technology has drastically increased the scope and speed with which intern architects design and document today's projects. Current digital tools require interns to learn how a building is put together much earlier in the design process. Furthermore, many aspects of design documentation that used to require hours of labor can now be completed in minutes. One hour in a firm today is much different than one hour in a firm when IDP was first created.[268]

Notes

1 1.1.1, for example *Spearin*.
2 For a discussion of, inter alia, a historiography of the separation of designers and constructors, see generally Scolari, M. *Oblique Drawing: A History of Anti-perspective*. Cambridge: MIT Press (2012), 139.
3 See generally Davis, S. D., & Prichard, R. *Risk Management, Insurance and Bonding for the Construction Industry*. Virginia: The Associated General Contractors of America (2000).
4 §2.2.
5 AIA Document B101™ 2007 Commentary. 2.
6 Brown, T. W., & Basham, J. R. "Building Information Modeling." In Michael T. Callahan (Ed.) *Shared Design*. Aspen Publishers (2011).
7 Friedlander, M. C., & Rowe, H. H. "A Practical Guide to New and Controversial Issues in AIA B101 – 2007." In J. Kent Holland (Ed.) *Risk Management for Design Professionals in a World of Change*. Ardent (2010).
8 §2.2.
9 §2.2.
10 §3.2.2.
11 §3.2.3.
12 §3.5.
13 §3.3.2.
14 *Howard v. Usiak*, 775 A. 2d 909 – Vt – Supreme Court 2001 citing *Coombs v. Beede*, 89 Me. 187, 36 A. 104, 89 Maine 187 (1896).
15 *Wells v. City of Vancouver*, 467 P.2d 292, 77 Wash. 2d 800, 11 Wa2d 800 (1970). 804.
16 741 S.W.2d 349 (Tex. 1987). 355.
17 Noteboom, L. J., & Hosfield, M. J. "Owner Claims against Construction Managers, Design Professionals, General Contractors, and Design-Builders." In R. F. Cushman (Ed.) *Proving and Pricing Construction Claims*. Aspen Publishers (1996), 12–16.
18 *Martin v. Sizemore*, 78 S.W.3d 249 (Tenn. Ct. App. 2001). 275.
19 http://cife.stanford.edu/mission. Accessed on December 7, 2014.
20 http://cife.stanford.edu/publications. Accessed on December 7, 2014.
21 Teicholz, P. M. *Technology Trends and Their Impact in the A/E/C Industry*. Stanford University, Center for Integrated Facility Engineering (1989).
22 See Mahoney, J. J., & Tatum, C. B. "Construction Site Applications of CAD." *Journal of Construction Engineering and Management* 120.3 (1994): 617–631.
23 Collier, E. B. *Four-Dimensional Modeling in Design and Construction*. Unpublished dissertation. Stanford University (1995).
24 Tatum, C. B., & Korman, T. "MEP Coordination in Building and Industrial Projects." CIFE Working Paper no. WP054, 03/1999. Stanford University, Center for Integrated Facility Engineering (1999).

25 Accessed on March 11, 2015. http://cife.stanford.edu/mission.
26 Gallaher, M. P., & Chapman, R. E. *Cost Analysis of Inadequate Interoperability in the US Capital Facilities Industry*. US Department of Commerce, Technology Administration, National Institute of Standards and Technology (2004).
27 Ibid. ES-2.
28 NIST, 6-3.
29 NIST.
30 Google Scholar citation search. Conducted on December 9, 2014.
31 The AGC released an updated *Contractors Guide Ed. 2* in 2010 which clarifies themes and adds additional content.
32 Associated General Contractors. *Contractors' Guide to BIM* (2006). 2.
33 Ibid. 2.
34 Ibid. 3.
35 Ibid. 5.
36 Ibid. 7.
37 AGC-BIM, 13.
38 Construction, McGraw-Hill. *The Business Value of BIM in North America: Multi-Year Trend Analysis and User Ratings (2007–2012)*. Bedford, MA: McGraw-Hill Construction (2012).
39 See, for example, Hanna, A., Boodai, F., & El Asmar, M. "The State of Practice of Building Information Modeling (BIM) in the Mechanical and Electrical Construction Industries." *Journal of Construction Engineering and Management* 139.10 (2013).
40 *Supra*, 3.1.4.
41 AGC-BIM, 14.
42 Ibid. 28.
43 Ibid. 25.
44 Ibid. 25.
45 Ibid. 26.
46 Ibid. 27.
47 Ibid. 28.
48 Accessed on March 11, 2015. http://www.agc.org/learn/education-training/building-information-modeling/cm-bim. Disclosure: The author has previously served as an approved instructor for AGC BIM coursework units.
49 Indiana University, Vice President for Capital Planning and Facilities, *BIM Guidelines & Standards for Architects, Engineers, and Contractors Sept. 10th 2009*.
50 Indiana University, *BIM Guidelines & Standards for Architects, Engineers, and Contractors Revised July 2nd, 2012*.
51 Ibid. 1.
52 Ibid. §1.2.
53 Ibid. §1.4.
54 Ibid. §1.5.
55 Ibid. §1.6.
56 "BIM Proficiency Matrix." Accessed on September 12, 2012. http://www.iu.edu/~vpcpf/consultant-contractor/standards/bim-standards.shtml.
57 Ibid. §2.2.
58 Accessed on March 11, 2015. http://www.iu.edu/~vpcpf/consultant-contractor/contracts/index.shtml.
59 Ibid.
60 §2.3.
61 §2.3.

62 Ashcraft, H. W. "IPD Teams: Creation, Organization and Management." (2011). Retrieved December 12, 2012. http://www.hansonbridgett.com/Publications/pdf/~/media/Files/Publications/IPD-Teams.pdf.

63 §2.4.

64 §2.5.

65 §4.2.6.

66 Ibid. §4.4.6.

67 Accessed on March 11, 2015. http://www.wbdg.org/resources/cobie.php.

68 IU-2012. §4.6.3.4.

69 *Supra* Chapter 2.6.

70 State of Wisconsin Department of Administration, Division of Facilities Development, *Building Information Modeling (BIM) Guidelines and Standards for Architects and Engineers* (Rev. 9/12).1

71 Ibid. 1.

72 State of Wisconsin Department of Administration, Division of Facilities Development. DOA-4519P (R09/12). Accessed on September 11, 2014. http://www.doa.state.wi.us/Documents/DFD/Forms/DOA-4519P%20AE%20Contract%20for%20Professional%20Svcs.pdf.

73 State of Wisconsin Department of Administration, Division of Facilities Development. DOA-4518P (R12/2013). Accessed on September 11, 2014. http://www.doa.state.wi.us/documents/DFD/Forms/DOA-4518P-AE-PPM.pdf.

74 Ibid. 4.C.

75 DOA-4193P (C01/14). Article 2.J.

76 §1.1.

77 §1.2, 1.3.

78 §1.2.

79 See, for example, §3.2.

80 §2.2.

81 US Dept. of Veterans Affairs (VA) *BIM Guide* v1.0. April 2010.

82 Ibid. §1.

83 Ibid. 7.1.

84 Ibid. 7.2.

85 Ibid. 7.3.

86 Ibid. 7.8.

87 Ibid. 7.9.

88 Ibid. 7.10.

89 Ibid. 2.2.

90 US Dept. of Veterans Affairs (VA) PG 18-15 Vol. B, Oct 2010, rev Aug 1, 2013. I-1., Pg 7.

91 Ibid. See, for example, III-B-2.a, Pg. 11.

92 VA A/E contract for design development. Article E, SPE 7. Accessed on September 9, 2014. http://www.cfm.va.gov/contract/ae.asp.

93 VA *Architectural Design Manual*. PG 18-10, §1.5 (August 1, 2014).

94 VAFR-BIM. §3.

95 Ibid. §3.

96 www.cfm.va.gov/. Accessed on September 9, 2014.

97 Reviewed in full *Infra* Section 3.1.4.1.

98 *Supra*. Chapter 3.1.3.1.

99 Ohio, Department of Administrative Services, General Services Division, State Architect's Office, "State of Ohio Building Information Modeling Protocol" (2010). 2.

100 Ibid. 8.
101 Ibid.
102 Ibid. 8.
103 Ohio Facilities Construction Commission. "Ohio Construction Reform." Accessed on September 9, 2013. http://ocr.ohio.gov/.
104 Ohio, 129th General Assembly, Amended Substitute House Bill Number 153 (2011). Accessed on September 9, 2013. http://ocr.ohio.gov/.
105 OH-BIM, 11.
106 Ibid.19.
107 Ibid.
108 Ibid
109 Ibid.
110 Ibid. 20.
111 Ibid.
112 Ibid. 21.
113 NYC-DDC. Accessed on September 14, 2014. http://www.nyc.gov/html/ddc/html/home/home.shtml.
114 Ibid.
115 RFP-6. PIN 8502012VP0028P-33P. Accessed on September 14, 2014. http://www.nyc.gov/html/ddc/html/home/home.shtml.
116 §1.9.
117 §6.3.3.
118 §7.2.2(a)6.
119 §6.3.9.
120 DDC-BIM Part II, Model Discrepancies, Pg. 17.
121 NYC Department of Design and Construction. RFP-20. PIN:8502012VP0008-27P. Accessed on September 14, 2014. http://www.nyc.gov/html/ddc/html/home/home.shtml.
122 NYC DDC Design Consultant Guide. C.11. Accessed on September 14, 2014. http://www.nyc.gov/html/ddc/html/home/home.shtml.
123 DDC-CG. 1.A.2.
124 DDC-BIM. Part 1. Pg. 8.
125 Ibid. 10.
126 Ibid.
127 Ibid. Part 2. Pg. 13.
128 Reviewed *Infra* 3.1.4.1.
129 DDC-BIM, 17.
130 Ibid. 20.
131 *National Building Information Modeling Standard, v2* (2012). Foreword.
132 Ibid. 1.
133 Ibid. 1.
134 Ibid. §5.2., 1.
135 Ibid. §5.2., 1.
136 Ibid. §5.5., 3.
137 United States National CAD Standard. Accessed on March 11, 2015. http://www.nationalcadstandard.org/ncs5/.
138 NCADS Overview. Accessed on March 11, 2015. http://www.nationalcadstandard.org/ncs5/about.php.
139 See generally Kolarevic, B. Ed. *Architecture in the Digital Age: Design and Manufacturing.* Taylor & Francis (2004). Addressing contemporary architectural practice in which

digital technologies are radically changing how buildings are conceived, designed and produced.

140 Gilmore, G. *The Death of Contract*. Columbus, OH: Ohio State University Press (1974), 2.

141 "AIA Revision Policy." American Institute of Architects. Accessed on September 4, 2013. http://www.aia.org/contractdocs/AIAS076676.

142 AIA categorization: C-Series *Other Agreements*; E-Series *Exhibits*; G-Series *Contract Administration/Project Management Forms*.

143 Sweet, Justin. *Sweet on Construction Industry Contracts: Major AIA Documents*. New York: Wiley (1996), 9.

144 See "Lack of Knowledge about BIM Standard Form Documents Widespread among Construction Professionals Responding to WPL Publishing's Latest Survey," ConstructionProNetwork. Accessed September 4, 2013. http://constructionpronet. com/Content_Free/2012-07-16CPC.aspx.

145 See Construction, McGraw-Hill. "The Business Value of BIM in North America: Multi-Year Trend Analysis and User Ratings (2007–2012)." (2012). Accessed September 4, 2013, http://images.autodesk.com/adsk/files/mhc_business_value_of_ bim_in_north_america_2007-2012_smr.pdf.

146 Sweet, *Sweet on Construction Industry Contracts*, 10.

147 The American Institute of Architects. Demkin, J. A., ed. *The Architect's Handbook of Professional Practice*. 14th ed. Hoboken, NJ: Wiley (2008).

148 The American Institute of Architects. Demkin, J. A., ed. *The Architect's Handbook of Professional Practice*. 13th ed. Hoboken, NJ: Wiley (2001), 376.

149 The American Institute of Architects. Haviland, D., ed. *The Architect's Handbook of Professional Practice*. 12th ed. Washington, DC: AIA Press (1994), 734.

150 AIA Comparison of AIA B-141 1987 and 1997 Editions. §2.4.2.1. Accessed on September 21, 2014. http://www.umich.edu/~cee431/aia_docs.htm.

151 The American Institute of Architects. Haviland, D., ed. *The Architect's Handbook of Professional Practice*. 11th ed. Washington, DC: AIA Press (1988).

152 The American Institute of Architects. *The Architect's Handbook of Professional Practice*. 10th ed. Washington, DC: AIA Press (1970). 12–13.

153 §1.1.1.

154 §1.1.

155 §1.2.1.

156 §1.2.3.

157 475 F.2d 601 (Ct. Cl. 1973); accord *D'Annunzio Broth. v. Transit Corp.*, 586 A.2d 301, 245 N.J. Super. 527 (Super. Ct. App. Div. 1991).

158 Ibid. 602.

159 760 F.2d 1288 (Fed. Cir. 1985).

160 *Supra* 2.3.

161 See NIST *infra* XX accord Becerik-Gerber, Burcin, et al. "Application Areas and Data Requirements for BIM-Enabled Facilities Management." *Journal of Construction Engineering and Management* 138.3 (2011): 431–442.

162 *Supra* IU-2012 at Section 3.1.3.1.

163 *Supra* DDC BIM at Section 3.1.3.5.

164 *Supra* 3.1.3.6.

165 1.2.2.

166 3.1.

167 BIMForum. *Level of Development Specification, Draft 1*. April 19, 2013.

168 3.2.2.2.
169 AIA G202, §3.2, see discussion *infra* 3.1.4.3.
170 §1.4.7.
171 See, for example, http://www.autodesk.com/products/bim-360-field/overview; http://www.latista.com/. Accessed on March 11, 2015.
172 E202™ (2008), §1.1.1.
173 §1.4.11, 1.4.12.
174 §1.1.2.
175 §1.3.1.
176 §1.3.3.
177 §1.3.3.
178 §1.4.4.
179 §1.4.8.
180 §1.4.9.
181 §1.4.10.
182 §2.3.
183 §2.1.
184 §1.4.8.
185 §3.1.
186 §3.2.
187 See, for example, IU-2012 *supra* 3.1.3.1, VAFR-BIM *supra* 3.1.3.2.
188 See *Guide, Instructions and Commentary to the 2013 AIA Digital Practice Documents.* Copyright © 2013 by the American Institute of Architects.
189 §3.2.1.
190 §3.2.2.
191 §3.2.3.
192 §3.4.2.
193 §4.1.
194 §4.5.
195 §4.5.2.
196 §4.5.3.
197 §4.6.
198 §4.8.2.
199 §4.8.3.
200 §4.10.
201 §2.1.1.
202 §2.1.2 and 2.1.3.
203 §2.2 and 2.3, respectively.
204 §1.1.
205 §1.3.
206 §2.4.1.
207 §2.6.1.
208 §2.5.2.4.
209 Article 3.
210 §3.1.1.
211. See note 71.
212 §3.2.
213 AIA Commentary.
214 ConsensusDocs. Accessed on September 4, 2013. https://www.consensusdocs.org/.

215 "ConsensusDocs Procedures." ConsensusDocs. Accessed on September 4, 2013. https://www.consensusdocs.org/procedures. See §4.1. Disclosure: At the writing of this book, the author was a volunteer participant in the ConsensusDocs 301 review and update working group.
216 ¶1.2.
217 ¶1.4.
218 ¶1.5.
219 ¶2.10.
220 ¶2.1.
221 ¶2.14.
222 ¶2.4.
223 ¶2.6.
224 §3.4.
225 AIA Document Commentary, B201 (2007). 11.
226 Ibid. §3.4.
227 Friedlander & Rowe.
228 ¶2.9.
229 Acret. James. *National Construction Law Manual.* 4th ed. BNi Publications, Inc. (2002).195.
230 For example, ¶4.3.3, 4.3.10, 5.3.
231 ¶2.2.
232 §2.8.
233 §2.7.
234 §1.7.
235 §1.10.
236 §1.9.
237 §2.11.
238 §3.1.
239 §3.2.1.
240 §3.2.2.
241 §3.2.3.
242 §3.2.5.
243 §4.1.
244 ¶4.3.2.
245 §4.3.5.
246 Ibid.
247 ¶4.3.12.
248 ¶4.3.13.
249 ¶4.3.16.
250 ¶4.3.9.
251 ¶4.3.8.
252 ¶4.3.11.
253 *Supra* XX.
254 IU (2012) §1.1.
255 NY DDC BIM (2012), Part One.
256 WI-BIM (2012) §1.1.
257 ¶4.3.21.
258 ¶4.3.22.
259 ¶5.1.

260 Lowe, R. H., & Muncey, J. M. "ConsensusDocs 301 BIM Addendum." *Construction Law* 29 (2009): 24.
261 ¶5.2.
262 ¶5.4.
263 998 P.2d 305, 140 Wash. 2d 568, 140 Wa. 2d 568 (2000). *Supra* 3.1.
264 ¶5.8.
265 ¶6.1.
266 ¶6.4.
267 For example, AIA E203 (2013), G201 (2013), G202 (2013); VAFR-BIM; NYC-DDC BIM.
268 http://blog.ncarb.org/2014/June/IDP-Proposals.aspx. Accessed on December 14, 2014.

References

Acret, James. *National Construction Law Manual*. 4th ed. BNi Publications, Inc. (2002).

American Institute of Architects. *The Architect's Handbook of Professional Practice*. 10th ed. Washington, DC: AIA Press (1970).

American Institute of Architects. Haviland, D., ed. *The Architect's Handbook of Professional Practice*. 11th ed. Washington, DC: AIA Press (1988).

American Institute of Architects. Haviland, D., ed. *The Architect's Handbook of Professional Practice*. 12th ed. Washington, DC: AIA Press (1994). 734.

American Institute of Architects. AIA Comparison of AIA B-141 1987 and 1997 Editions. §2.4.2.1. Accessed September 21, 2014. http://www.umich.edu/~cee431/aia_docs.htm.

American Institute of Architects. AIA Document Commentary, B201 (2007). 11.

American Institute of Architects. A201 (2007) *General Conditions of the Contract for Construction*.

American Institute of Architects. B101 (2007) *Standard Form of Agreement between Owner and Architect*.

American Institute of Architects. C132 (2009) *Standard Form of Agreement Between Owner and Construction Manager as Advisor*.

American Institute of Architects. Demkin, J. A., ed. *The Architect's Handbook of Professional Practice*. 13th ed. Hoboken: Wiley (2001). 376.

American Institute of Architects. Demkin, J. A., ed. *The Architect's Handbook of Professional Practice*. 14th ed. Hoboken: Wiley (2008).

American Institute of Architects. Document B101™ 2007 Commentary.

American Institute of Architects. E202 (2008) *Building Information Modeling Protocol Exhibit*.

American Institute of Architects. E203 (2013) *Building Information Modeling and Digital Data Exhibit*.

American Institute of Architects. G201 (2013) *Project Digital Data Protocol Form*.

American Institute of Architects. G202 (2013) *Project Building Information Modeling Protocol Form*.

American Institute of Architects. *Guide, Instructions and Commentary to the 2013 AIA Digital Practice Documents*.

Ashcraft, Howard W. "IPD Teams: Creation, Organization and Management." *Hanson Bridgett, IPD/BIM*, (2011). Retrieved December 12, 2012. http://www.hansonbridgett.com/Publications/pdf/~/media/Files/Publications/IPD-Teams.pdf.

Associated General Contractors. *AGC Contractors' Guide to BIM* (2006).

Becerik-Gerber, Burcin, et al. "Application Areas and Data Requirements for BIM-Enabled Facilities Management." *Journal of Construction Engineering and Management* 138.3 (2011): 431–442.

BIMForum. *Level of Development Specification, Draft 1.* April 19, 2013.

Brown, T. W., & Basham, J. R. "Building Information Modeling." In Michael T. Callahan (Ed.) *Shared Design.* Aspen Publishers (2011).

Center for Integrated Facility Engineering (CIFE), Stanford University. Accessed on December 7, 2014. http://cife.stanford.edu/mission.

Collier, E. B. *Four-Dimensional Modeling in Design and Construction.* Unpublished dissertation. Stanford University (1995).

ConsensusDocs. 301 *Building Information Modeling (BIM) Addendum* (2008).

Construction Operations Building Information Exchange (COBie). Accessed on December 1, 2014. http://www.wbdg.org/resources/cobie.php.

Davis, S. D., & Prichard, R. *Risk Management, Insurance and Bonding for the Construction Industry.* Virginia: The Associated General Contractors of America (2000).

Friedlander, M. C., & Rowe, H. H. "A Practical Guide to New and Controversial Issues in AIA B101 – 2007." In J. Kent Holland (Ed.) *Risk Management for Design Professionals in a World of Change.* Ardent (2010).

Gallaher, M. P., & Chapman, R. E. *Cost Analysis of Inadequate Interoperability in the US Capital Facilities Industry.* US Department of Commerce, Technology Administration, National Institute of Standards and Technology (2004).

Gilmore, Grant. *The Death of Contract.* Columbus, OH: Ohio State University Press (1974).

Hanna, A., Boodai, F., & El Asmar, M. "The State of Practice of Building Information Modeling (BIM) in the Mechanical and Electrical Construction Industries." *Journal of Construction Engineering and Management* (2013).

Indiana University, Vice President for Capital Planning and Facilities, *BIM Guidelines & Standards for Architects, Engineers, and Contractors.* Sept. 10th, 2009 (Revised July 2, 2012).

Indiana University. "BIM Proficiency Matrix." Accessed September 12, 2012. http://www.iu.edu/~vpcpf/consultant-contractor/standards/bim-standards.shtml.

Indiana University. Consultant Contractor contracts. Accessed September 12, 2012. http://www.iu.edu/~vpcpf/consultant-contractor/contracts/index.shtml.

Kolarevic, Branko, Ed. *Architecture in the Digital Age: Design and Manufacturing.* Taylor & Francis (2004).

Lowe, R. H., & Muncey, J. M. "ConsensusDocs 301 BIM addendum." *Construction Law* 29 (2009): 17.

Mahoney, J. J., & Tatum, C. B. "Construction Site Applications of CAD." *Journal of Construction Engineering and Management* 120.3 (1994): 617–631.

McGraw-Hill. *The Business Value of BIM in North America: Multi-Year Trend Analysis and User Ratings (2007–2012).* Bedford, MA: McGraw-Hill Construction (2012).

National Building Information Modeling Standard, v2 (2012).

New York City Department of Design and Construction. *Design Consultant Guide.* C.11. Accessed on September 14, 2014. http://www.nyc.gov/html/ddc/html/home/home.shtml.

New York City Department of Design and Construction. RFP-6. PIN 8502012VP0028P-33P. Accessed on September 14, 2014. http://www.nyc.gov/html/ddc/html/home/home.shtml.

New York City Department of Design and Construction. RFP-20. PIN:8502012VP0008-27P. Accessed on September 14, 2014. http://www.nyc.gov/html/ddc/html/home/home.shtml.

Noteboom, L. J. & Hosfield, M. J. "Owner Claims against Construction Managers, Design Professionals, General Contractors, and Design-Builders." In R. F. Cushman (Ed.) *Proving and Pricing Construction Claims.* Aspen Publishers (1996).

NYC *Department of Design and Construction BIM Guidelines* (2012). Accessed on September 14, 2014. http://www.nyc.gov/html/ddc/html/home/home.shtml.

Ohio. 129th General Assembly, Amended Substitute House Bill Number 153 (2011). Accessed on September 9, 2013. http://ocr.ohio.gov/.

Ohio, Department of Administrative Services, General Services Division, State Architect's Office. *State of Ohio Building Information Modeling Protocol* (2010).

Ohio Facilities Construction Commission."Ohio Construction Reform." Accessed on September 9, 2013. http://ocr.ohio.gov/.

Scolari, M. *Oblique Drawing: A History of Anti-perspective.* Cambridge: MIT Press (2012).

State of Wisconsin Department of Administration, Division of Facilities Development. *Building Information Modeling (BIM) Guidelines and Standards for Architects and Engineers* (Rev. 9/12).1.

State of Wisconsin Department of Administration, Division of Facilities Development. DOA-4518P (R12/2013). Accessed on September 11, 2014. http://www.doa.state. wi.us/documents/DFD/Forms/DOA-4518P-AE-PPM.pdf.

State of Wisconsin Department of Administration, Division of Facilities Development. DOA-4519P (R09/12). Accessed on September 11, 2014. http://www.doa.state. wi.us/Documents/DFD/Forms/DOA-4519P%20AE%20Contract%20for%20 Professional%20Svcs.pdf.

Sweet, Justin. *Sweet on Construction Industry Contracts: Major AIA Documents.* New York: Wiley (1996).

Tatum, B. C., & Korman, T. "MEP Coordination in Building and Industrial Projects." CIFE Working Paper. Stanford University, Center for Integrated Facility Engineering (1999).

Teicholz, P. M. *Technology Trends and Their Impact in the A/E/C Industry.* Stanford University, Center for Integrated Facility Engineering (1989).

United States Department of Veterans Affairs. *BIM Guide v1.0.* April 2010.

United States Department of Veterans Affairs. PG 18-15 Vol. B, Oct 2010, rev Aug 1, 2013. I-1.

United States Department of Veterans Affairs. VA A/E contract for design development. Article E, SPE 7. Accessed on September 9, 2014. http://www.cfm.va.gov/contract/ ae.asp.

United States Department of Veterans Affairs. VA *Architectural Design Manual.* PG 18-10, §1.5 (August 1, 2014).

United States National CAD Standard. Accessed at: http://www.nationalcadstandard.org/ ncs5/.

List of cases

Coombs v. *Beede*, 89 Me. 187, 36 A. 104, 89 Maine 187 (1896)

D'Annuzio Broth. v. *Transit Corp.*, 586 A.2d 301, 245 N.J. Super. 527 (Super. Ct. App. Div. 1991)

Fortec Constructors v. *United States*, 760 F.2d 1288 (Fed. Cir. 1985)

Howard v. *Usiak*, 775 A. 2d 909, 172 Vt. 227 (2001)

MA Mortenson Co. v. *Timberline Software Corporation*, 998 P.2d 305, 140 Wash. 2d 568, 140 Wa. 2d 568 (2000)

Martin v. *Sizemore*, 78 S.W.3d 249 (Tenn. Ct. App. 2001)

Melody Home Mfg. Co. v. *Barnes*, 741 S.W.2d 349 (Tex. 1987)

Merando, Inc. v. *United States*, 475 F.2d 603, 201 Ct. Cl. 28 (Ct. Cl. 1973)

Wells v. *City of Vancouver*, 467 P.2d 292, 77 Wash. 2d 800, 11 Wa2d 800 (1970)

4 Legal issues and claims considerations

The purpose of this chapter is to explore BIM/VDC in the context of legal issues and consider notions of claims. Thus, to start – a disclaimer. The present categorization might be considered by some to be an actor in search of a play. As of the researching of this chapter the collection of published court decisions dealing specifically with BIM/VDC is thin. A lone standout is MA *Mortenson Co. v. Timberline Software Corporation*.[1] In *Mortenson*, a general contractor purchased off-the-shelf bid preparation software. While using the software to prepare a bid for a project, two of the contractor's employees encountered an "Abort" error message approximately five to seven times during the course of the final day's work.[2] After each instance, the contractor's employees checked their work in the software and it appeared accurate to them. Work proceeded and the general contractor submitted a bid, but that bid was later discovered to be two million dollars too low as a result of a bug in the software. The appellate court's analysis in the case was limited to questions of law concerning the licensing agreement in the packaging and instruction manuals of the software and any resulting contract formation or alteration between the parties. The court in *Mortenson* did not, nor were experts employed, to examine or offer specific commentary on standard of care or workmanlike performance regarding a contractor's use of BIM/VDC tools or processes. The court in *Mortenson* held the software maker's liability for errors in the software was limited to the license fee paid for the right to use the software.

Though there is no reported case, the BIM/VDC claim most typically referenced is that as reported in *Architectural Record* in 2011.[3] While the names and specific details remain hidden, the broad strokes include a university life-sciences building where the architect and his MEP consultant used BIM to coordinate above-ceiling. While the BIM was well coordinated to a high degree of detail, it also required a specific sequencing to achieve the fit. The particular sequencing requirements were not relayed to the contractor. As a result, once the contractor was at roughly 70 percent completion of install they ran out of space. The contractor brought suit against the owner, who in turn sued the architect which ultimately resulted in the MEP engineer being included. As reported, the parties settled for "millions of dollars" for fear that a jury would be unable to comprehend the issues.[4] Aside from this reported claim, the author here has participated in the analysis of disputes involving BIM/VDC components.

Returning to published cases, in addition to *Mortenson* the US courts have produced other decisions with regard to technology's impact on construction. For example, there are cases involving construction scheduling technologies. A survey might include *RW Vaught Co.* v. *FD Rich Co.*[5] The case involved a claim brought by the mechanical subcontractor Vaught against the prime contractor Rich for unpaid contract amounts, wherein Rich had, inter alia, utilized a computerized scheduling program to schedule a project consisting of several buildings and a heating plant on a US Army base. Rich prepared several updates to the computer-derived schedule for the heating plant resulting in at least three different completion dates for that particular portion of the work. Rich had then submitted schedule print-outs to each subcontractor as a guide for the subcontractor's own schedule preparation. With respect to the computer-derived schedule the trial court found "the evidence in this case clearly demonstrated that this electronic brain (located in Connecticut) received a great deal of bad information from the defendant's employees in Missouri" and further categorized the completion dates generated by the computer as "arbitrary" and theoretical aspirations.[6] The US Court of Appeals for the Eighth Circuit affirmed in favor of Vaught on this point, noting, "the record indicates . . . that the prime contractor fed erroneous data into the computer."[7] In the parlance of many modern BIM/VDC conversations, "garbage in, garbage out."

This is all to say that the following categorization of BIM and VDC legal issues and potential claims is being done (as the chapter title indicates) with consideration. Consideration of published court reports, indeed the lack thereof. Consideration of current industry literature and trade-press reports concerning BIM legal and claims issues. Consideration of how the courts have dealt with technological changes in the construction industry. Consideration of the direct professional BIM/VDC claim and mediation experience of the author. Consideration of decades-long technological natural selection versus meteor-induced technological eradication.

Practically, the rubric for analysis is separated into two issue "sieves." The first issue sieve has been broadly termed "professional/contract" and leads off with a discussion of responsible control. The second issue sieve, "technical," explores issues such as 2D–3D conversions, data loss, and so on. Needless to say, neither sieve is intended to block the passage or influence of concepts from one to the other. A basic matrix showing where recent form contract documents address general legal concerns and issues associated with BIM/VDC can be found in Table 4.1.

4.1 Professional/contract – responsible control

The United States NBIMS definition for BIM includes "collaboration" as a fundamental characteristic of BIM technologies and processes.[8] Standing in contrast against this general premise of collaboration is a current reality comprised of a multifaceted design and construction ecosystem – professional, legal, insurance – that clearly articulates distinction and separation with very carefully defined rules

Table 4.1 General legal issues in BIM/VDC form documents

Form contract	Standard of care	Responsible control of design	Spearin warranty	Model development, use, and reliance	Legal status of the model	2D–3D conversions	Interoperability	Software version control	Data loss, archiving and preservation	Copyright and intellectual property	Duty to inform
ConsensusDocs 301 (2008)	¶5.4	¶1.4	¶1.5	Articles 4; 5	¶1.10; 2.3; 4.3.9; 4.3.11		¶4.3.19	¶4.3.18	¶4.3.7; 4.3.8	Article 6	§1.9
AIA E202 (2008)				Articles 3; 4					§2.4.4	§2.2	§2.1
AIA E203 (2013)				Articles 4					Digital data: §3.3; 3.5; BIM: §4.8.3; 4.8.4 §2.3	§2.1	
AIA G201 (2013)				All	Article 2		§1.7				
AIA G202 (2013)									§1.7; 1.8		§3.1.2

of engagement between all parties. Thus, where BIM as a highly collaborative endeavor bumps up against the daily realities of a design and construction project, the industry must address, amongst other issues, the legal concept of "responsible control." Prior to examining the industry's initial responses to responsible control in BIM/VDC form documents, some brief context on the concept generally.

In describing the means by which an architect assumes responsible control the National Council of Architectural Registration Boards' (NCARB) *Rules of Conduct* read, in part, "An architect may sign and seal technical submissions only if the technical submissions were: (i) prepared by the architect; (ii) prepared by persons under the architect's responsible control." Responsible control is later defined in the same section:

> that amount of control over and detailed professional knowledge of the content of technical submissions during their preparation as is ordinarily exercised by architects applying the required professional standard of care. Reviewing, or reviewing and correcting, technical submissions after they have been prepared by others does not constitute the exercise of responsible control because the reviewer has neither control over nor detailed knowledge of the content of such submissions throughout their preparation.[9]

Similar to NCARB, the National Society of Professional Engineers' defines "responsible charge" for the engineer, in part, as "that degree of control an engineer is required to exercise over engineering decisions made personally or by others over which the engineer provides supervisory direction and control authority."[10]

Following the lead of NCARB and its stated mission to provide tools and model procedures for individual US jurisdictions, most states therefore require the architect of record to be in responsible charge of a design.[11] In the Commonwealth of Pennsylvania, for example, relevant portions of the state Code concerning the use of the architect's seal of registration dictate, "An architect may not impress the seal or stamp, or knowingly permit it to be impressed or affixed, on drawings, specifications or other design documents which were not prepared by the architect or under his direct supervision."[12] In Colorado statutes define the architect's responsible control as "that amount of control over and detailed knowledge of the content of plans, designs, drawings, specifications, and reports during their preparation as is ordinarily exercised by a licensed architect applying the required standard of care."[13]

While in no way BIM specific, *Wynner v. Buxton*[14] offers an exploration of responsible control. The case involved the owner of a self-service gas station who entered into a contract with a steel frame building contractor to engineer, obtain permits, erect, and provide supervision of the erection of the gas station. The owner brought suit against a licensed engineer who was acting as a consultant to the building contractor. The owner alleged that the engineer's negligent preparation of the plans and specifications resulted in buildings that leaked, the failure of fuel lines between storage tanks and delivery equipment, asphalt paving failure as a result of improperly compacted soil, and improper site drainage. At trial, the engineer moved for summary judgment, contending that his preparation of certain drawings, and signature upon the same, was specifically related to structural steel calculations for the buildings only. While his signature was also upon certain other drawings describing the entire work, his drawings specifically excluded details related to civil engineering aspects of the work such as soil compaction, mechanical systems, and site drainage which the owner was alleging deficient. Likewise, the engineer's notes specifically indicated the need for separate permits regarding piping and fuel lines. The trial court agreed and granted the engineer's second motion for summary judgment.

Upon appeal, the court in *Wynner* affirmed the trial court's decision holding that the engineer's responsibility for the civil engineering plans specific to his scope of work for the steel buildings did not also make him responsible for all the plans covering all aspects of the project. Further, the failure of those civil engineering plans related to his scope of work to contain all details of all aspects of the work did not render them defective. The court noted, for example, that under then current and applicable building codes, those additional details might be supplied by a licensed contractor in preparing shop or field drawings.[15] The court also held that being in responsible control of design plans did not also impose a duty to supervise execution of the work.[16] The *Wynner* decision illustrates an instance where one professional's responsible control of a portion of a design limited his responsibilities for an overall design.

Questions concerning how technology might impact responsible control have been broadly considered in some of the earliest CAD research.[17] With the highly collaborative and multi-participant, multi-contribution nature of a BIM/VDC process the topic has drawn new attention. Fundamentally, potential issues could arise from a BIM/VDC process that entails the creation and aggregation of any number of relevant models and/or model contributions, at varying degrees of development, and by any number of potential contributors – not all of whom may be licensed professionals. Architects, engineers, specialty consultants, subcontractors, building product manufacturers, out-sourced service providers – each might be contributing content (with that content perhaps iterative of previous, unrelated content) to a BIM/VDC-enabled project. The architect/engineer's ability to remain in responsible control of each design contribution to a BIM/VDC project, which cumulatively constitute the design, could present challenges if not managed.

Contractual provisions between an architect and his usual professional consultants (e.g. structural, mechanical) would likely put those professionals in responsible control of their respective scopes of service and place upon them a duty to perform their services (which might include BIM/VDC content) in accordance with the applicable standard of care.[18] But what of content produced by non-professionals and then utilized by either the architect or professional consultants? As others have questioned, do manufacturers, vendors, perhaps even standards committees developing BIM object specifications become "designers" in a collaborative process?[19] Returning to the NCARB definition from the outset of this section, could an architect exhibit control over, and have detailed professional knowledge of, BIM content during its preparation by these other non-professionals?

To address such questions both the ConsensusDocs 301 *BIM Addendum* (explicitly) and AIA E203 (2013) *Building Information Modeling and Digital Data Exhibit* (implicitly) address responsible control.

Commentary by Lowe and Muncey (legal counsel involved in the drafting of the ConsensusDocs 301) indicates that the drafters of the *Addendum* adopted the position that there was no restructuring of typical contractual relationships in utilizing the *Addendum* and that the architect/engineer remains the person in responsible charge of the design.[20] Accordingly, Section 1.4 of the *Addendum* states "Nothing in this Addendum shall relieve the Architect/Engineer from its obligation, nor diminish the role of the Architect/Engineer, as the person responsible for and in charge of the design of the Project."

The combination of the 301's status as an addendum to governing and/or affiliated contracts each with their own liability flow-down provisions, along with explicit language related to responsible control, are methods intended to allay the "anxiety" felt by many early BIM implementers as noted by Lowe and Muncey, and others.[21]

Similar to the ConsensusDocs 301, the AIA E203 (2013) is not a stand-alone document and is intended to be incorporated into any agreement(s) with, "Project Participants that may develop or make use of Digital Data on the Project."[22] While the E203 (2013) does not make an explicit declaration regarding responsible control of design, given its provenance as an AIA document one can infer the AIA's intention/expectation that the architect is in responsible control of the design. Likewise, by default, the E203 does assign the architect other responsibilities including: preparing and distributing digital data protocols, managing and maintaining any centralized digital document management solution intended to be used on the project, and preparing and distributing modeling protocols.[23]

The industry documents above have taken steps to address responsible control of design on BIM/VDC-enabled projects. As the standard of care and workmanlike performance continuously evolve, so too will the functional means to address responsible control in these documents, as well as others yet to be created. However, while form documents might enable certain types of industry transformation at the project level, additional systemic changes will also be necessary. As Ashcraft has noted, statutes defining responsible control and professional licensing must also adapt to the realities of actual digital practice.[24] Such efforts may already be underway. In June 2014, NCARB posted commentary related to proposed changes to their Intern Development Program for training and preparing architects for licensure. Part of the proposed changes includes a 33 percent reduction in the amount of apprenticeship hours currently required by the intern program. An internal NCARB task force evaluating the current program included the following assessment as one of three self-diagnosed issues with the current IPD structure:

2   Technology has drastically increased the scope and speed with which intern architects design and document today's projects. Current digital tools require interns to learn how a building is put together much earlier in the design process. Furthermore, many aspects of design documentation that used to require hours of labor can now be completed in minutes. One hour in a firm today is much different than one hour in a firm when IDP was first created.[25]

While these comments represent a proposal that as of researching this book still required vetting through NCARB's member boards and, ultimately, the NCARB board of directors, they unequivocally state an appreciation of technology's actual impact on architectural practice. And though not addressing responsible control per se, they do concern licensing policy for emerging architects who upon registration will deliver services applying a technologically influenced standard of care.

4.2 Professional/contract – model development, use, and reliance

In a construction claim that includes allegations of deficient contract drawings the following questions might be asked. "Are the documents suitable for their intended purpose, whether it be pricing, bidding, or construction?" And, "Do the plans and specifications adequately define quantity, configuration, and quality?" Parties to such a claim will likely have opposing responses, but there is nonetheless the implication that standards and customs for each purpose do exist. That is, as a project progresses through the typical schematic design (SD), design development (DD), and construction documents (CDs) phases of a project, the corresponding drawings and documents at each phase are generally understood to progress cumulatively, building up in detail, and to deliver a corresponding level of reliability. For example, what were simply two lines describing a wall's overall thickness in a small scale floor plan during SD, have, by the CD phase, taken on many more lines to express much more graphic detail (e.g. visualization of insulation, sheathing, weatherproofing, etc.) and corresponding qualitative information (e.g. performance) in the form of specification. Both the designer and the contractor understand the former to be an incomplete, yet an appropriately reliable, indication of what's to come, whereas the latter presents a more exact accounting of what is specifically required for bidding the work.[26]

An example of how courts have addressed the preparation of progressive design drawings that relied upon the work product of others in traditional 2D workflows can be found in *Taylor v. DeLosso*.[27] In *Taylor* an owner plaintiff contracted separately with a licensed land surveyor for preparation of a site survey, and an architect (#1) (then currently employed by company A). Architect #1(A) in responsible control of the design subsequently utilized and certified the survey as part of a site plan submission for a zoning variance on behalf of the owner. After obtaining initial zoning approval, the owner contracted again with the same architect #1 (then currently employed by company B) to prepare a revised site plan that reflected additional project scope. While architect #1(B) later affirmed that he did not use a site survey in preparation of the revised site plan while working at company B, the revision was also approved by the local zoning board.

During the course of construction the contractor was forced to halt work when it was discovered that the location of a 30" diameter maple tree was shown incorrectly on the revised site plan as prepared by architect #1(B). Plaintiff then hired another architect, #2, to begin anew. As a result of his review of the existing documentation, and a visit to the job site, architect #2 discovered that both the original survey and the revised site plan by architect #1(B) had each located the maple tree incorrectly – approximately 11' to 14' from its true geospatial location on the site.

As part of its analysis, the *Taylor* court found that "reliance" as defined by the applicable state laws, and in conjunction with reliance provisions of the contract between the plaintiff and architect #1 (A and B) (which were in turn modeled after an industry form contract, namely the AIA B-141), provided architect #1 as the person in responsible charge of the design with an "unqualified right" to incorporate the work of the licensed surveyor.[28]

Customary notions of design drawing progression, use, and reliability that accompany traditional 2D contract documents have garnered significant attention with BIM/VDC workflows. What should a BIM, or more accurately the discrete objects within an overall BIM, actually look like (graphics) and contain (information) during the various stages of design and construction? When and how does a wall progress from simply describing its overall thickness to actually containing more individual details about the materials of its assembly? With the inherent high degree of collaboration of BIM/VDC tools and workflows, how does the author of a piece of BIM content indicate to what extent others can rely on their work at a given point in the design and construction cycle? A working resolution to these questions and others like them has given birth to another acronym, "LOD," or level of development.

As part of the emerging lingua franca in the US AEC industry, LOD is a specific designation applied to BIM content that signifies and articulates the degree of completeness of the content, as well as the corresponding use and reliance provisions that consumers/receivers of the BIM content are entitled (or barred) from applying at different stages in the design and construction process. An overly simplistic illustration might be, "Project participant EA contributed X to the BIM. It can be used for K at milestone 2. It can be used for D and T at milestone 3. It may never be used for L."

The AIA E202 (2008) was the first industry form contract document in the US to define LOD and introduce its attendant levels of progression, 100, 200, 300, 400, and 500, into the greater BIM/VDC vernacular.[29] The LOD concept was carried over into the revised E203 (2013) and, more specifically, the associated AIA G202 (2013) *Project Building Information Modeling Protocol.* Accordingly, the E203 (2013) and G202 (2013) are reviewed below.

The E203 (2013) defines LOD as "the minimum dimensional, spatial, quantitative, qualitative, and other data included in a Model Element to support the

Authorized Uses associated with such LOD."[30] Thus, for example, LOD 200: "The Model Element is graphically represented within the Model as a generic system, object, or assembly with approximate quantities, size, shape, location, and orientation. Non-graphic information may also be attached to the Model Element."[31] "Generic" and "approximate" are the operative words to describe model content typically associated with the earlier (e.g. SD/DD) phases of design. Likewise, LOD 300 (roughly analogous to construction documents) reads, "The Model Element is graphically represented within the Model as a *specific* system, object or assembly in terms of quantity, size, shape, location and orientation. Non-graphic information may also be attached to the Model Elements" (emphasis added).[32] LODs are cumulative, such that LOD 300 includes and builds upon the characteristics of LODs 200 and 100.

For ease of conceptualizing, most practitioners correlate LODs to progressive project phases, that is, SD (100) through as-built construction (500). As a result of this correlation many in the AEC industry initially interpreted this to mean that an entire BIM achieved a given LOD at the conclusion of a project phase. However, the realities of the design, construction, and modeling process can result in variable LODs per object or system within a BIM at a given point in time. As such, a BIM is never collectively, say, a "LOD 200 BIM." The G202 (2013) addressed this issue in part by changing the descriptive header for content requirements at each LOD. Whereas the E202 (2008) read, "Model Content Requirements" the G202 (2013) reads, "Model *Element* Content Requirements" (emphasis added). This initial perception has also been self-correcting in the industry. Likewise, an emerging industry guideline, the *LOD Specification* developed by the Associated General Contractors BIMForum, seeks to further clarify best-practice LOD usage. Building upon the AIA definitions, the *LOD Specification* has specifically developed LOD 350 for cross-trade coordination.[33] As reported by Post through her interviews of contributors to the *LOD Specification*, most litigation generally occurs between concepts of LOD 300 and LOD 400.[34]

In addition to defining the progressive geometric and non-graphic characteristics of Model Elements, each LOD identifies and defines various "Authorized Uses" of model content at the given LOD. Authorized uses seek to provide Model Element Authors (MEA) with the ability to convey and control expectations and use of the content by others. For example, the authorized use for cost estimating at LOD 200 states, "The model may be used to develop cost estimates based on the approximate data provided and quantitative estimating techniques (e.g. volume and quantity of elements or type of system selected.)"[35] With such a definition the author of BIM content is effectively saying, for example, "I've modeled this HVAC duct run in approximately the right place and so you may use it to conceptually estimate linear feet of material."

Types of authorized use are dependent upon LOD. For example LOD 100 to 400 uses include: "Analysis," "Cost Estimating," "Schedule," and "Other Authorized Uses" as defined by the parties. However, as testament to continued refinement of the standard of care with respect to BIM/VDC, the G202 (2013) has added a new authorized use of "Coordination" for LODs 200 to 400.

The coordination definition is specific to each LOD and increases in specificity. For example, LOD 200 use for coordination states: "The Model Element may be used for coordination with other Model Elements in terms of its size, location, and clearance to other Model Elements." LOD 400 authorizes a Model Element to be coordinated to the same terms, with the addition of "fabrication, installation and detailed operation issues."[36]

The addition of an authorized use regarding coordination provides for, amongst other possible benefits, relief from coordinating to unnecessary or cumbersome dimensional tolerances too early in a project, as well as a structure for intelligently incorporating long-term facilities and maintenance use requirements into a project at appropriate intervals.

In another example of standard of care refinement, LOD 500, while conveying essentially the same requirements regarding as-built model elements as set forth in the E202 (2008), uses new clarifying language, "The Model Element is a *field verified* representation in terms of size, shape, location, quantity, and orientation. Non-graphic information may also be attached to the Model Elements" (emphasis added).[37] Further, for LOD 500 the general usage definition used in the E202 (2008) related to using content for maintaining, altering, and adding to the project as granted by any applicable license(s) has been removed in the G202. This was likely done to coordinate and reinforce Section 2.3 of the E203 (2013) which expanded a party's usage rights beyond design and construction to include "using, maintaining, altering and adding to the Project." This language matches Section 7.3 of the AIA's B101 owner–architect agreement (2007) that provides owners with a nonexclusive license to use an architect's instruments of service.

With LOD definitions and authorized uses defined, the G202 (2013) also describes reliance.[38] In a slight, but pointed change from the E202 (2008) the definition for reliance is built around project milestones as opposed to project phases. This was likely done to reinforce the above noted fact that the model element progression is not necessarily intended to be perfectly analogous to traditional project phasing. Accordingly, Section 3.1.1 reads:

> At any particular Project milestone, a Project Participant may rely on the accuracy and completeness of a Model Element only to the extent consistent with the minimum data required for the Model Element's LOD for that Project milestone as identified below in the Model Element Table, even if the content of a specific Model Element includes data that exceeded the minimum data required for the identified LOD.[39]

For example, one might envision a scenario in which a mechanical engineer has downloaded BIM content from a specific equipment manufacturer for a particular type of pump during early phase DD. While the content as developed by the manufacturer may contain a high level of geometric and informational detail, the rules of engagement dictate that it can only be utilized and relied upon by others consistent with the LOD assigned for the given milestone.

With reliance expectations established, Section 3.1.2 then establishes a duty with respect to identification of conflict stating, in part:

> Where conflicts are found in the model, regardless of the phase of the Project, or LOD, the Project Participant that identifies the conflict shall promptly notify the Model Element Authors and the Project Participant . . . responsible for Model Management. Upon such notification, the Model Element Author(s) shall act promptly to evaluate, mitigate and resolve the conflict in accordance with the processes established in Sec. 1.7.7, if applicable.

As with the E202 (2008), the G202 (2013) utilizes the "Model Element Table" as a visual control matrix for project participants to organize BIM content development, coordinate modeling work across disciplines, and establish reliance.[40] Functionally, model elements are assigned an MEA, and applicable LOD(s) across project milestones. Whereas the E202 (2008) defaulted to the CSI Uniformat™ classification system for organizing model elements, the G202 (2013) sample document leaves determination of the classification system to project participants.

In another adjustment from the E202, the G202 changes the definition for MEA (that term now defined in the E203 (2013)). In the original E202 (2008), MEA is defined in Section 1.2.3 as "the party responsible for *developing the content* of a specific Model Element to the LOD required for a particular phase of the Project" (emphasis added).[41] As recast in the E203 (2013) the MEA is now defined as "the entity (or individual) responsible for *managing and coordinating* the development of a specific Model Element to the LOD required for an identified Project milestone, *regardless of who is responsible for providing the content* in the Model Element" (emphasis added).[42] The AIA indicates that MEA assignments do not alter who is in responsible control for those given parts of the BIM/design, and that MEA assignments may change as the project transitions from design to construction, for example design steel to fabricator steel.[43]

Like AIA documents, the ConsensusDocs 301 (2008) *BIM Addendum* also provides mechanisms for organizing model content development and assigning responsibility and reliability in BIM/VDC workflows. As a baseline, the *Addendum* defines a "Design Model" as a model that has reached a level of completion typically associated with 2D construction drawings.[44] Likewise a "Construction Model" is defined, in part, to contain information equivalent to typical shop drawings.[45] Procedurally, Section 4 requires the project participants to establish and formalize a tactical document – a "BIM Execution Plan" (BEP) – which specifies BIM roles, responsibilities and requirements. Foremost amongst these requirements is an identification of what models/content are required, and who will have responsibility for those models/content.[46] While the ConsensusDocs 301 does not include a matrix for organizing that effort as does the AIA model element table, it has been suggested that the *Addendum* might interface with AIA documents, generally.[47] Once assigned responsibility for model content in the BEP, each party takes responsibility for any contribution that it makes to a model, or a contribution by another project participant in privity with that party and of a lower tier than that party.[48] In short, one who contributes BIM content, takes responsibility for that content.

In addition to assigning model content responsibilities, the *Addendum* also requires project participants to establish the dimensional accuracy of the model itself. Paragraph 4.3.11 Contributor's Dimensional Accuracy Representation requires the parties to choose one, and only one, of four available choices related to the dimensional accuracy of BIM content. Each of the first three choices in Paragraph 4.3.11 progressively limits the dimensional reliability of BIM. Prior to making their single selection, any choice is qualified as being limited to other parties to the governing contract, in accordance with the standard of care applicable to the contributor, and effective at the time the model has parity with traditional 2D CDs.[49] The fourth choice enables the project team to draft their own language regarding model dimensional accuracy.

With both content responsibility and dimensional accuracy representation established, the *Addendum* also addresses reliance in Section 5.3:

> To the Extent that any or all Design Models are included as Contract Documents, Project Participants may rely upon the accuracy of information in those Design Models; provided, however, that regardless of whether any Design Models are included as Contract Documents, the selection in Section 4.3.11 shall control a Project Participant's right to rely on the dimensional accuracy of a Contribution or Model.

As both BIM/VDC technology and industry standards advance, specific procedures for the development, use, and reliability of BIM content will also be refined. At present, if one considers the example claims questions posed at the start of this section from a BIM/VDC standard of care perspective – for example "Are the documents suitable for their intended purpose, whether it be pricing, bidding, or construction?" or, "Do the plans and specifications adequately define quantity, configuration, and quality?" – it would appear that current industry structures for BIM development, use, and reliability could support such an analysis.

4.3 Professional/contract – legal status of the model

The preceding sections considered responsible control, and model development, use, and reliance in the context of BIM/VDC. Accordingly, it appears prudent to now ask, just what is the legal status of a BIM? More to the point – is BIM a contract document – or better yet – do permitting agencies accept BIM? An analysis into such questions is launched from a review of typical and customary definitions of drawings and contract documents.

The AIA A201 (2007) *General Conditions of the Contract for Construction* defines "Drawings" as "the graphic and pictorial portions of the Contract Documents showing the design, location, and dimensions of the Work, generally including plans, elevations, sections, details, schedules and diagrams."[50] The A201 (2007) also defines (for the first time in AIA documents) "Instruments of Services" as "representations, in any medium of expression now known or later developed, of the tangible and intangible creative work performed by the Architects . . . [and] may include, without limitation, studies, surveys, models, sketches, drawings, specifications, and other similar materials."[51] The AIA's commentary sheet for the A201 (2007) states that the instruments of service term and definition were included to acknowledge that "technological advances such as CAD have and will continue to have an impact on the architect's services and the manner in which they are provided."[52] Thus it would appear that while neither explicitly defining BIM in the A201 (2007) nor including BIM as a constituent part of contract documents, the AIA does, understandably, wish to protect the architect's rights and ownership of content produced in a technologically-enabled design practice. Similarly, the AIA B101 (2007) *Agreement Between Owner and Architect*, while making mention of digital models as perhaps being part of the SD phase submission, and listing BIM as a potential additional service, provides no definition of BIM.[53]

In light of the typical contract definitions above, a turn towards the AIA E-series documents for digital practice. All of the E-series documents are standalone documents and they must be attached to an existing agreement. A possible scenario in the early adoption of BIM/VDC may have had an existing agreement, say the AIA B101 (2007) *Agreement Between Owner and Architect*, executed by the parties, with that governing contract perhaps having a BIM-specific exhibit, such as the E202 (2008), or otherwise. However, most existing agreements had, in almost all cases, neither altered their definition of "Contract Documents" to include BIM, nor separately defined BIM. In such a scenario it would appear then that a BIM is not a contract document. As noted by Aschraft, the BIM status might be considered an, "'accommodation document' that can be used, but not relied upon."[54]

Nor do the E-series documents themselves define BIM as being a contract document. Though removed in the current G202 (2013) *Project Building Information Modeling Protocol* definition for LOD 300, the initial E202 (2008) did include "Construction" as an authorized use of BIM content at LOD 300. There "Construction" was defined as "Suitable for the generation of traditional construction documents and shop drawings."[55] While neither capitalizing nor defining CDs or shop drawings elsewhere in the exhibit, with this definition the E202 nonetheless appeared to have acknowledged the technological reality that BIM supports the direct extraction and creation of 2D documentation.[56] In certain instances BIM is the 2D drawing and the 2D drawing is BIM insofar as bi-directional associativity allows an end user to directly edit either the model object, or the 2D projected drawing view of that object, with the end result being the same alteration to both the model and the drawing. Hence, if drawings are defined as pictorial and graphic representations of the work and such a drawing might be indistinguishable from BIM (i.e. bi-directional associativity being merely the transient medium of digital 1's and 0's encapsulating the content in that moment before it changes from one state to the next), then under what specific circumstances, can one include or exclude BIM from contract documents?

As previously discussed, the Indiana University *BIM Guidelines* (IU-2012) may present one such scenario.[57] Neither IU's publically available governing contracts for architects, nor their operating guidelines for contractors appear to designate BIM as a contract document. However, Section 1.1.1 of IU-2012 – which is incorporated by reference to each of the aforementioned – specifically requires that, "All drawings, schedules, simulations, and services required for assessment, review, bidding and construction shall be extractions from this model [a BIM]." It would appear here then, that one cannot create a design drawing (contract document) without it being an extraction from the BIM. Similar to IU-2012, the State of Wisconsin *BIM Guidelines* also require all design drawings to be extractions from models.[58]

In contrast to AIA documents, the ConsensusDOCS 301 *BIM Addendum* enables BIM to be considered a contract document. Paragraph 2.3 states, "Contract Documents, as defined in the Governing Contract, is modified to include all Design Models unless otherwise specified in the BIM Execution Plan." Additional clauses provide means to limit BIM as a contract document. For example, the BEP, as a required amendment to the *Addendum*, requires the project team to also define any design models that do not constitute contract documents.[59] While this provision allows for a measured approach to excluding certain design-side BIM content from the contract documents definition, additional sections offer more sweeping control.

Paragraph 4.3.11 Contributor's Dimensional Accuracy Representation requires that project participants choose one, and only one, of four available choices related to the dimensional accuracy of BIM content. Disregarding bidirectional associativity, this clause asks, "What counts – drawing(s) or the BIM?" The language in the three drafted choices progressively restricts the dimensional reliability of BIM, with the fourth "other" choice leaving the door open for project stakeholders to draft their own language for BIM reliability. The first check box is the most progressive in terms of BIM/VDC stating that a content contributor "represents that the dimensions in its Contribution to a Model are accurate and take precedence over the dimensions called out in the Drawings or Inferred from the Drawings." The second choice, as one might logically anticipate, narrows BIM reliance slightly dictating that each contributor "represents that the dimensions in its Contribution to a Model are accurate to the extent that the BIM Execution Plan specifies dimensions to be accurate, and all other dimensions must be retrieved from the drawings." The third drafted choice is the least BIM/VDC progressive stating, "Contributors make no representation with respect to the dimensional accuracy of the Contributor's Contribution to a Model. A model can be used for reference only and all dimensions must be retrieved from the Drawings." To the dismay of many early BIM/VDC proponents, the third option was often the choice selected by project participants in the early ascent of the adoption curve.

In addition to contract-based issues for considering the legal status of a BIM, the global AEC industry has also been investigating the use of BIM/VDC for regulatory code compliance and permitting. For example, an industry consortium project being administered by FIATECH called the *AutoCodes Project* seeks to streamline a more efficient and uniform code-checking process based on BIM submissions.[60] To date, the Building and Construction Authority (BCA) of Singapore is considered the global leader in such advancements. Beginning in 2010, nine regulatory agencies in Singapore accepted architectural BIMs for approvals, followed in 2011 with the acceptance of MEP models. As of 2011 more than 200 projects had made BIM e-submissions to the BCA.[61] While it appears that no US-based jurisdictions have yet to allow e-submissions of BIMs, publically available BIM guidelines, such as the NYC Department of Design and Construction's, have noted the potential benefits of BIM-enabled code checking compliance.[62]

As with notions of responsible control and BIM content development and reliance, the legal status of a BIM will continue to evolve. Some BIM guidelines have already recognized the best practice value in requiring all design drawings to be extractions from a BIM. Likewise, the ConsensusDocs 301 has enabled a BIM to be deemed a contract document. As technology continues to advance along with broader education of public officials, it also seems likely that BIM code-compliance capabilities such as those currently being utilized in Singapore will spread. The trend would appear to be moving away from relegating BIM/VDC to third class status as "accommodation documents" and towards placing them squarely on par with traditional definitions for drawings and contract documents.

4.4 Technical – 2D–3D conversion

The AEC industry was quick to realize that achieving the "M" in BIM, whose 3D geometry would in turn support a VDC spatial coordination process, may, in many cases, require the need to convert 2D construction drawings into 3D models in the first place. This was driven by the practical contradiction that many contractors were (and remain) interested in the potential cost savings and risk mitigation of clash detection offered by a 3D BIM/VDC process, whereas many design professionals do (and for the near future will continue) to deliver their contract documents as flat, 2D physically printed hard copies. As previously noted, the AGC BIM – which can be counted amongst the inaugural class of popular industry guideline documents – not only addressed the issue of 2D–3D conversion broadly, but also offered a specific financial estimate of the effort involved.[63]

2D–3D conversions are considered here initially in terms of the technical procedures and the resultant fidelity (primarily geometric) between a 2D original and its 3D copy. In overly simplistic terms, what steps were taken, what buttons were clicked, is the copy faithful to the original, and so on? Reviewing technical procedures can support analysis of potential claims questions such as, did the conversion reveal errors or omissions in the 2D drawings, or did the conversion process create an error or omission in the BIM that could have consequences if the parties deem the BIM a controlling contract document? Questions around 2D–3D conversions might be within, or alongside, other questions around standard of care, responsible control, use and reliability, interoperability, or other sub-topics discussed herein. It is noted that this discussion does not presuppose a different potential for errors or increased liability to the parties than currently exists in the tracing and iterative use of purely 2D–2D document workflows. Accordingly, a summary discussion of current basic technical methods for 2D–3D conversions follows.

An original 2D contract document may be brought into a BIM authoring tool as a digital underlay where the 2D line work of the original is traced and given a third dimension in the BIM application. An underlay may be achieved by various technical methods including linking (i.e. "X-ref"/externally referenced), or importing. Without disregarding that digital underlay methods such as these each entail their own specific technical characteristics, the basic idea is roughly analogous to sliding an original document beneath a piece of onion skin onto which it will be traced.

In geometric terms, the x and y coordinates of the contract document are traced over, with the BIM authoring tool enabling the operator to simultaneously provide a z coordinate which has been established by, most likely, referencing other coordinated drawings within the contract documents set (such as an elevation or section). It is at this point that another disclaimer must be inserted insofar as depending upon situational variables: there are any number of permutations to this basic workflow that may affect the correlation between an original contract drawing and its corresponding BIM trace.

For example, currently, if an original contract document is brought into a BIM authoring tool as a resolution-dependent, 2D raster format (e.g. JPEG) and the user wishes to trace a series of lines symbolizing a wall on the original, the user will need to approximate the physical start and end points of those lines and corroborate them with any associated dimensions provided elsewhere on the same or different contract documents. This is because a BIM authoring tool cannot locate precise, mathematical coordinate system control points or nodes of the lines in the x,y work plane to "snap" to on raster-based drawings. Raster drawings are based on pixel resolution where the closer one zooms in, the less clear the image becomes. This is not to suggest that high precision and accuracy are required in every circumstance (they may not be), nor that the desired precision and accuracy could not be achieved through some other means or methods in model creation. Nor does this marginalize the fact that some BIM authoring tools do provide for features that seek to streamline and add degrees of accuracy in utilizing raster documents for underlay procedures. Likewise, this does not disregard the existence of raster-to-vector conversion tools which could offer some degree of flexibility in working with raster documents in the first instance. It is merely meant to note that potential for error or misinterpretation in the 3D tracing of 2D raster-based documents could develop, much the way error or misinterpretation could occur in traditional 2D–2D based workflows.

In contrast, if the original contract document being traced is vector-based (e.g. DWG) the user will have the ability to "snap" to the coordinate-based control points or nodes of those original lines, thereby suggesting greater parity between the original line and the trace line. Additionally, certain BIM authoring tools such as Autodesk® Revit® currently allow a user to "lock" relationships between original lines and their trace such that a change to the original results in a corresponding change in the copy upon saving and refreshing each.[64] While such workflows and features may increase continuity and accuracy of a 2D to 3D trace, any number of scenarios might be imagined which could upset the applecart.

Suppose a 3D trace was "locked" to the 2D original. The 2D design original was then superseded by a revision. Given the "locked" relationship, the change in the original should automatically be reflected in the trace. However, was the revision transition effectively managed? Was the original digital file – the one locked to the BIM – simply overwritten, or was the revision named as a new drawing and inserted into the contract set? Even with available link management and file path control tools in a BIM authoring tool, does the availability of those tools guarantee that the correct drawing is being traced? Organizationally, are there discrepancies within the project team where certain groups are abiding by the most current, newly named 2D drawing, yet others are abiding by the BIM trace thought to represent the single source of truth? Furthermore, there is a possibility (though perhaps more rare amongst the digitally inclined?) that one performing the 2D to 3D conversion could *visually* reference hard-copy printed versions of the original 2D contract documents and replicate those drawings as a BIM without undertaking any of the import-underlay-trace workflows described above or otherwise envisioned. A figure-drawing exercise of sorts.

Potential scenarios abound that, depending upon choices or communication between the authors of the 2D original and the BIM trace, might strengthen or weaken the credibility and utility of the final BIM. Several BIM guidelines have taken steps to avoid unnecessary translation between 2D and 3D design drawings by requiring that all drawings be extractions from a BIM, and attempting to prohibit the use of disconnected 2D files and drawings.[65] Again, at the expense of repetition, there is nothing to suggest a different potential for errors or increased liability to the parties in a 2D–3D conversion than currently exists in the tracing and iterative use of purely 2D–2D document conversions.

Having considered the technical aspects of 2D–3D conversions, what does the process offer generally in terms of exposing ambiguities, errors or omissions, and so on in the contract documents themselves? The end product of a 2D–3D conversion inherently presents an opportunity to consider BIM/VDC variables within the context of typical construction claims questions. The basic visualization and coordination features of BIM/VDC tools and processes can provide efficiency and insight in evaluating, "Are there errors or omissions in the documents (now revealed or clarified by BIM)?", "Are the documents (now BIM) suitable for their intended purpose whether it be pricing, bidding, or construction?", "Are the documents (now BIM) reasonably complete, coordinated, and internally consistent?", "Do the plans (now BIM) and specifications (perhaps now also BIM) adequately define quantity, configuration, and quality?" or "Has adequate interdisciplinary design coordination been effectively and timely implemented?" The questions that might arise in potential BIM/VDC claims may ultimately look nothing like their historical ancestors from a paper-based process, but as with most paradigm changes, imitation is usually the first evolutionary step.

4.5 Technical – interoperability

Engaging the term 'interoperability' might best take flight from a short historical runway. That history reveals that over the course of several millennia the preparation and utility of traditional 2D drawings has changed very little. Contemporary architects are engaged in techniques and deliverables much the same as at least their 15th century BC forbearers which, as documented in the work of Vitruvius in *De Architectura*, identified the plan, elevation, and section as the three types of drawings necessary to represent architecture.[66] These plan, elevation, and section drawings function primarily via geometry (e.g. points, lines, and planes) and standardized symbology (e.g. a specific cross-hatch, irregularly spaced dots, dashes, and triangles). Lines on a page, combined in certain ways, convey geometric messages, say the thickness of a cross-section through an aluminum window frame, or the length of a concrete foundation wall.

These lines and graphic symbols on the flat page visually represent physical entities, but they don't in and of themselves necessarily convey all additional relevant data or information – say the alloy of an aluminum extrusion or the specific strength characteristics of concrete. This is achieved through other means including text notation, schedules, specifications, professional custom. Nor do the lines on the page understand how they interact with other lines on the page. The lines representing a door placed within other lines representing a wall are simply a group of lines arranged in a particular configuration. Lines themselves are very good for describing the geometry of the entity, but not much else in the complex endeavor that is designing, building, and ultimately operating a building.

Given such a history, a logical initial outcome of the transition from preparing drawings by hand to drafting with 2D CAD programs was improving upon the means to describe geometric entities in plan, section, and elevation. 2D CAD programs developed as entity-based software focused on geometry and symbology that mimicked the processes of hand drafting. Thus, irrespective of any efficiencies that CAD may have brought to the process of document production, the end result – the digital or printed version of a 2D CAD drawing – was still primarily lines on a screen or page.

Accordingly, as multiple software vendors began to bring various mass-market CAD solutions to the AEC market in the 1980s, the ability to share information between them was, to a certain degree, a matter of enabling the lines from one program to be visibly seen and printed in another. Vendors worked to enable file sharing and created exchange methods (e.g. Drawing eXchange Format [DXF]) between their proprietary file extensions, but the line describing the concrete wall did not necessarily know it was a concrete wall as it moved from one program to the next. It was simply a line in any CAD program. At the expense of marginalizing what must have been (and doubtless remains) a significant effort for early CAD vendors, as long as the lines from one application could be visually seen in, and printed from, another application, then custom and culture would do the rest.

At the risk of tying to summarize a systemic transition, the process of exchanging lines worked efficiently enough until the core development of design technology and CAD programs began to shift from entity-based towards object-based.[67] With object-based design software the focus is on capturing relevant data and relationships about the entity, not merely its geometry. The former *entity* is now understood beyond simply its *x,y* coordinates as a domain-specific, information rich *object*. An interior partition wall for example, "knows" that it is a particular type of wall, connected to other walls, each of which have characteristics, such as fire ratings, or doors placed within their limits, and that collectively those walls might enclose a space with a given volume and function. The professional custom of interpreting flat geometry and symbology is now supported and enhanced by an extensive database of information, with 3D capabilities, that might be useful for any number of design or construction analyses or upstream/downstream purposes. Hence the acronym for BIM itself – Building *Information* Modeling – information being the operative word. While the low hanging fruit has been the "M," it's the "I" that holds the most promise. As noted by Bernstein and Pittman, a useful analogy might be Microsoft® Word® (for purposes here – "entity") versus Excel® ("object"). The former being static and isolated (text [lines] on the page), the latter a series of operable relationships (cells [walls] connected to other cells [walls] with operable/calculable purpose and meaning [room volume for HVAC load calculations]). Exploring design and construction scenarios in Word® presents challenges. Excel®, by contrast is purpose-built to support scenario exploration.[68]

As a result of the shift from entity-based to object-based design software the focus on sharing likewise shifted from the concept of exchange to that of interoperability. Data and information must now, as the term reveals, be interoperable, not merely visually exchangeable for viewing and printing. The ability to view 3D geometry does not qualify as true interoperability in and of itself. A wall in one system must "know" that it is also a wall in another system, such that either might want to calculate thermal performance, or manage a facility's paint coating schedule in a third or fourth application, both of which also understand the object to be a wall. To be interoperable in a market-driven economy some Latin must present itself to communicate between the various proprietary object-based systems.

At present, that non-proprietary method is the Industry Foundation Classes (IFC) as developed by buildingSMART (formerly the International Alliance for Interoperability). As stated on the buildingSMART website, the IFC data model is "a common data schema that makes it possible to hold and exchange data between different proprietary software applications. The data schema comprises information covering the many disciplines that contribute to a building project throughout its lifecycle: from conception, through design, construction and operation to refurbishment or demolition."[69]

With their charter as international, neutral, non-profit, buildingSMART serves as the independent organization for the creation of an interoperable

BIM data exchange schema and the certifying body for any proprietary BIM software wishing to achieve IFC certification. A matrix with the current IFC Certification status of any BIM vendors seeking certification is posted on the buildingSMART website.[70] While IFC represents a tremendous effort in achieving true interoperability between and amongst the myriad of proprietary systems in the design–construct–operate continuum, as with any large undertaking some early limitations have been identified (e.g. not all geometric constraints describing shapes being exchangeable required improvements for details needed for fabrication and manufacturing). While the exchange from one software application to another is a first branch on the IFC tree of interoperability, the end-goal ultimately seeks to improve and streamline the associated workflows and processes.[71]

Most current BIM guidelines address interoperability requirements to varying degrees. For example, whereas the IU *BIM Guidelines* addresses "interoperability" more broadly,[72] the US Department of Veterans Affairs' *BIM Guide* specifically requires use of BIM authoring software that is compliant with the then most current IFC format.[73] The State of Wisconsin *BIM Guidelines* requires the design team to submit an IFC deliverable at project close-out that reflects contractor recorded changes.[74] The ConsensusDocs 301 *BIM Addendum* requires the parties to account more generally for interoperability within a tactical BEP.[75]

As noted above, true interoperability is more than the exchange of 3D geometry between software applications. Accordingly, parties will want to ensure that interoperability has been properly addressed where data exported from a BIM is intended to interface, that is, be interoperable, with additional design, construction, and business software systems. However, fundamentals of geometric exchange remain relevant. Similar to the ways in which parties will want to be informed regarding 2D–3D conversions and any possible errors or omissions in the end-product BIM, parties will also want to ensure that necessary steps have been taken to ensure a complete exchange of 3D geometric information. While market leading project review tools such as Autodesk® Navisworks® allow for the compiling of numerous file extensions, without the installation of so-called "object enablers," it is possible that all geometry from a contribution file may not be visible.[76] Parties cannot coordinate what they cannot see, or did not know to exist.

4.6 Technical – software version control

It is a given of the modern condition that technological innovation occurs at break-neck speed. The pace of development and release within the AEC industry is no exception. As a result, larger design and construction projects may see several annual release versions of the BIM/VDC solutions they are utilizing over the course of their design and construction schedules. A project that starts in version X, could end in version Z.4. Software vendors typically provide documentation detailing any potential impacts that upgrades to newer versions, either annual or intermittent, will have to existing file structures, and so on. Typically, any impacts have to do with one-time performance degradation while new features are applied, after which the end user experiences the benefit of enhancements.

Likewise, some vendors offer eligible customers the ability to access previous versions, thus allowing multiple projects at various stages of design or construction to maintain use of their current version without the need to utilize only the most current version. Some BIM authoring software requires all files that seek to communicate with one another to be on the same version, for example a BIM on version 2008 cannot interact with a BIM on version 2009. Given the various personalities of project participants and their distinct business policies and budget cycles for technology it is critical that project teams anticipate technological changes over the expected duration of their projects and plan accordingly. Current industry form documents from both the AIA and ConsensusDocs provide for such planning,[77] as do most BIM guidelines.[78]

4.7 Technical – data loss, and data archiving and preservation

While the topics overlap, data loss is distinguished here from data archiving and preservation. Data loss is concerned with issues including on-going access, monitoring, and back-up capabilities for BIM/VDC data during the course of a project. Data archiving and preservation is capturing and maintaining any contractually required BIM/VDC milestone data for the record both during and after the conclusion of a project.

As noted by Ashcraft, insuring against data loss requires close attention, including considering purely economic losses.[79] (Recall that privity of contract and economic losses have shown themselves to be contentious topics.[80]) Given the inherent risk of data access, monitoring, and loss prevention in highly collaborative BIM/VDC workflows, early form contract BIM documents from both the AIA and ConsensusDocs contain clauses concerning data loss.

Section 3 of the ConsensusDocs 301 *BIM Addendum* is dedicated to information management and places a responsibility on the information manger to perform, or procure, various functions concerning BIM data loss. Examples include: encryption methods, back-up and restoration, monitoring system logs, and remedies for documented vulnerabilities.[81] The 301 includes additional measures such as the mutual waiver of consequential damages,[82] and flow-down provisions for risk allocation regarding use of and access to models by all project participants.[83] From an insurance perspective, the 301 also enables the parties to establish minimums for valuable papers and records coverage with regard to a project participant's BIM contributions.[84]

The AIA E203 (2013) contains provisions with respect to the maintenance, storage, and archiving of digital data in general, as well as BIM data management. Whereas the precursor E202 (2008) was intended for BIM purposes, the expanded E203 (2013) allows the parties to establish whether or not a centralized electronic document management system will be utilized. If so, the parties are subsequently required to establish any relevant protocols in the accompanying G201 (2013) *Protocol*.[85] Within the G201, the parties will specify procedures and requirements including on-going maintenance, storage, and archiving.[86] Similarly, the parties will establish BIM exchange, storage, maintenance, and archiving intentions within the E203, and then formalize more specific protocols within the accompanying G203.[87]

Much like current contract drawing sets are preserved at applicable project milestones, so too must BIM deliverables be preserved. As seen above, early BIM form documents from both the AIA and ConsensusDocs require parties to establish the requirements and procedures for BIM content archiving and preservation. While neither document prescribes specific archiving or preservation requirements, at a minimum parties should be prepared to archive BIM data at the conclusion of typical design phases (SD, DD, and CDs), as well as appropriate construction milestones (trade zone sign-off, and so on). Care should be taken to ensure that all constituent BIM contributions are preserved.

4.8 Technical – copyright and intellectual property

Culturally, copyright and intellectual property rights have garnered additional attention with the increased ease of access to, and transmission of, data. As noted throughout this work, the highly collaborative, multi-participant, multi-contribution structure of BIM/VDC processes only sharpens questions around, "Who owns what?" Although not AEC specific, the courts have specifically addressed digital modeling, derivative works, and copyright protections. In *Meshwerks, Inc.* v. *Toyota Motor Sales USA, Inc.*,[88] the Tenth Circuit considered if wireframe mesh models of automobile designs resulted in sufficiently original designs that warranted copyright protections. The wireframe mesh models were produced by digitizing actual physical cars and then post-processing or "sculpting" those digital models to produce exact replicas for marketing campaigns.[89] The Tenth Circuit affirmed the District Court decision that the uncontested facts of the case did not show the wireframe mesh models to have included anything original of their own. Indeed, the contractual intent of the models was to exactly replicate an image of the cars, and thus the models did not deserve copyright protection. As suggested by Brunka, the use of BIM software for building renovation work may constitute a derivate work of an original copyright holder to the building design, but one that deserves exception to charges of copyright infringement.[90]

Before highlighting copyright protections in industry form documents for BIM/VDC and publically available BIM guidelines, a review of typical contract definitions and expectations. With the issuance of the A201 (2007) *General Conditions of the Contract for Construction*, the AIA defined (for the first time) "Instruments of Services" as "representations, in any medium of expression now known or later developed, of the tangible and intangible creative work performed by the Architects . . . [and] may include, without limitation, studies, surveys, models, sketches, drawings, specifications, and other similar materials."[91] The AIA's commentary sheet for the A201 (2007) states that the term Instruments of Service was included and defined so as to acknowledge that "technological advances such as computer-aided design, have and will continue to have an impact on the architect's services and the manner in which they are provided."[92] It follows that the B101 *2007 Standard Form of Agreement Between Owner and Architect* states that the architect and his consultants shall be deemed the authors and owners of their respective instruments of service, retaining all rights, including copyrights, and granting to owners a nonexclusive license for their use.[93]

From a BIM/VDC perspective, rights of data ownership in the AIA E203 (2013) are dealt with in fundamentally the same manner as the predecessor E202 (2008). That is, a party transmitting digital data conveys no ownership rights to the receiving party. Ownership rights are controlled by the governing agreements.[94] Similarly, while the rights of any party receiving data are limited to using, modifying, and further transmitting the data only as it relates to the project, the E203 (2013) does expand applicable uses beyond just designing and constructing the project. In a likely response to commentary by owners wanting to use BIM deliverables for operations and maintenance post-construction, Section 2.3 expands usage rights beyond design and construction to include "using, maintaining, altering and adding to the Project." This language matches that found in Section 7.3 of the AIA's B101 *2007 Owner Architect Agreement* that provides owners with the above noted nonexclusive license to use an architect's instruments of service.

Along with data ownership, Article 2 of the E203 (2013) provides clauses controlling transmission of digital data. A party transmitting digital data is required to warrant that it owns the copyright on the data it is transmitting, "or otherwise has permission to transmit the Digital Data for its use on the Project."[95] Separate clauses in Sections 2.2 and 2.2.1 establish similar transmission duties with respect to "Confidential Digital Data" which is defined as "Digital Data containing confidential or business proprietary information" that has been clearly marked as such.[96]

Similar to the AIA, the ConsensusDocs 301 *BIM Addendum* includes explicit terms regarding copyright and intellectual property rights. For example, the *Addendum* requires that each party warrant to other parties to the governing contract that it either holds the copyright to all of its contributions, or that it is licensed or otherwise authorized to use the contribution. Furthermore, each party agrees to indemnify and hold harmless the other in the event of third-party claims "arising out of or relating to, claims or demands relating to infringement or alleged infringement."[97]

Likewise, Paragraph 6.1 states that "each Party grants to the other Party or Parties to the Governing Contract (a) a limited, non-exclusive license to reproduce, distribute, display and otherwise use that Party's Contributions" for the purposes of the project. Sublicensing language also allows for similar use of contributions by project participants in affiliated contracts. The net effect is to allow all project participants to governing and affiliated contracts to utilize models and contributions in the delivery of their contractual obligations on the project.

While the preceding provisions clarify project participants' use of contributions during the course of designing, constructing and otherwise delivering the project, the owner's use of the full design model post-construction is controlled by the governing contract between the owner and architect.[98] As noted throughout this book, given the interest of owners to utilize BIM deliverables for facilities management, governing contracts are increasing, and most BIM guidelines are now being drafted accordingly. For example, the Ohio *BIM Protocol* includes specific language regarding model ownership, "Ownership of the Model: BIM models and facility data developed for the project are the property of the project owner. The owner may make use of this data as allowed under the laws of the State of Ohio for electronic data and contract documents."[99] Additional, non BIM-specific language within the OH-BIM states, in part:

> State Law currently stipulates that the project Owner holds title to the Contract Documents. With changes in technology, SAO interpreted this to include digital data as early as the April 2007 Standard Requirements. However, we recognize the distinction between the concepts of "instruments of service" and "intellectual property of the design," and identify additional compensation if the project becomes a prototype constructed on multiple sites.[100]

4.9 Summary

This chapter explored possible legal issues and potential claims relative to BIM/VDC. The analysis started with a disclaimer that both reported case law and popular reports of BIM/VDC claims/disputes are sparse. Nonetheless, the literature reveals that major issues such as responsible control, legal status of the model, and so on are being considered from a BIM/VDC perspective. In the case of the ConsensusDocs 301, issues like *Spearin* warranties and standard of care are explicitly clarified. While issues like 2D–3D conversions may present an environment for claims to emerge, there is nothing to suggest those conversions will have a different potential for errors or increased liability to the parties than currently exists in the tracing and iterative use of pure 2D–2D documents. In terms of copyright and intellectual property the court in *Meshwerks, Inc. v. Toyota Motor Sales USA, Inc.* was forced to address digital modeling, generally. Specific to BIM/VDC, form documents from the AIA and ConsensusDocs address data ownership. Likewise, many owner BIM guidelines are explicit in their ownership of BIM deliverables with an eye towards long-term facility management and renovation work. Whether or not BIM/VDC claims will increase in number or type was not a concern. This chapter was, as the title indicates, a consideration of these issues for the purpose of critical thinking.

Notes

1 970 P.2d 803, 93 Wash. App. 819 (Ct. App. 1999).
2 Ibid. 806.
3 Post, N. "BIM Lawsuit Offers Cautionary Tale." (2011). Accessed July 14, 2014. archrecord.construction.com/news/2011/05/110519-BIM-Lawsuit-1.asp.
4 Ibid.
5 439 F.2d 895 (Court of Appeals 1971).
6 Ibid. 900.
7 Ibid. 900.

8 *Supra* 2.1.
9 National Council of Architectural Registration Boards. *2014–2015 Rules of Conduct*. Rev. July 2014. Rule 5.2.
10 National Society of Professional Engineers. *Position Statement No. 1745—Responsible Charge*. Adopted: April 2005; latest revision: July 2010.
11 Walker, S. G., Holderness, R. A., & Butler, S. D. *State-By-State Guide to Architect, Engineer, and Contractor Licensing*. Aspen Law & Business (1999).
12 49 Pa. Code §9.142.
13 C.R.S. 12-25-302 (7) (2013).
14 97 Cal. App. 3d 166, 158 Cal. Rptr. 587 (Ct. App. 1979).
15 Ibid. 175.
16 Ibid. 176.
17 See Mahoney, J. J. *Barriers to CADD in the ACE Industry*. Stanford. CIFE. Technical Report 23 (1990).
18 See, for example, AIA C401-2007.
19 Ashcraft, H. W. "Building Information Modeling: A Framework for Collaboration." *Construction Law* 28 (2008): 183.
20 Lowe, R. H., & Muncey, J. M. "ConsensusDOCS 301 BIM Addendum." *Construction Law* 29 (2009): 18.
21 Lowe and Muncey, 18, referencing Larson, D. A., & Golden, K. A. "Entering the Brave, New World: An Introduction to Contracting for Building Information Modeling." *William Mitchell Law Review* 34 (2007): 75.
22 AIA E203 §1.2.
23 AIA E203 §3.2.1, §4.5.2, §3.5.3, respectively.
24 Ashcraft. 184.
25 http://blog.ncarb.org/2014/June/IDP-Proposals.aspx. Accessed on December 14, 2014.
26 See AIA-B101 (2007).
27 725 A.2d 51, 319 N.J. Super. 174, 319 N.J. (Super. Ct. App. Div. 1999).
28 Ibid. 56.
29 The concept of "Level of *Detail*" (emphasis added) was originally developed by a software vendor, Vico Software. Subsequently, the AIA California Council on IPD explored the notion of Level of *Detail* which was then incorporated by the AIA into the E202 (2008) as Level of Development. See, for example, http://www.aecbytes.com/viewpoint/2013/issue_68.html. Accessed on March 11, 2015.
30 E203 (2013) §1.4.4.
31 E203 (2013) §2.3.1.
32 AIA E203 (2013) §2.4.1.
33 BIMForum. *Level of Development Specification, Draft 1*. April 19, 2013. 9.
34 Post, N. "Help Coming for BIM Users." *Engineering News Record*. May 20, 2013. 15.
35 G202 (2013) §2.3.2.2.
36 G202 (2013) §2.5.2.4.
37 G202 (2013) §2.6.1.
38 Article 3.
39 §3.1.1.
40 §3.2.
41 AIA E202 §1.2.4.
42 AIA E203 §1.4.6.
43 AIA E203 *Guide*. 12.
44 301. ¶2.6.
45 301. ¶2.2.
46 §4.3.2.

47 Lowe and Muncey. Note 6.
48 §5.1.
49 ¶4.3.11.
50 Section 1.1.5.
51 Section 1.1.7.
52 AIA Document Commentary A201-2007, 6.
53 AIA B101 §3.2.5 and §4.1.6, respectively.
54 Ashcraft. 176.
55 Section 3.4.2.1.
56 *Supra* 2.5.
57 *Supra* 3.1.3.1.
58 *Supra* 3.1.3.4.
59 ¶4.3.3.
60 http://www.fiatech.org/index.php/projects/active-projects/162-active-projects/
    projects-management/593-automated-code-plan-checking-tool-proof-of-concept.
    Accessed on December 15, 2014.
61 *Build Smart*. Singapore Building and Construction Authority. Issue 9. December
    2011. 4.
62 DDC-BIM. Code Validation. 13.
63 *Supra* 3.1.2.3.
64 http://help.autodesk.com/view/RVT/2015/ENU/. Accessed on December 16, 2014.
65 See, for example, IU-20012 *supra* 3.1.3.1, WI-BIM *supra* 3.1.3.4.
66 Translated in 1914 as *Ten Books on Architecture* by Morris H. Morgan, Ph.D, LL.D.
    Harvard University. Chapter 2, Para. 2. Full text, Project Gutenberg. Accessed on
    September 6, 2014. http://www.gutenberg.org/ebooks/20239. There is scholarly
    debate over translation of the term for "section." See generally Scolari, M. *Oblique
    Drawing: A History of Anti-perspective*. Cambridge: MIT Press (2012). 164. Note 1.
67 See generally Eastman, C., Teicholz, P., Sacks, R., & Liston, K. *BIM Handbook: A
    Guide to Building Information Modeling for Owners, Managers, Designers, Engineers and
    Contractors*. 2nd ed. Wiley (2011).
68 Bernstein, P. G., & Pittman, J. H. "Barriers to the Adoption of Building Information
    Modeling in the Building Industry." *Autodesk Building Solutions* (2004).
69 http://www.buildingsmart.org/standards/ifc/model-industry-foundation-classes-
    ifc. Accessed September 6, 2014
70 http://www.buildingsmart.org/certification/currently-certified-software-prod-
    ucts. Accessed September 6, 2014.
71 Eastman. 118.
72 IU-2012 §1.4.
73 VAFR-BIM §9.
74 WI-BIM. 3.7.
75 ConsensusDocs 301. 4.3.19.
76 http://help.autodesk.com/view/NAV/2014/ENU/. Accessed on December 16, 2014.
77 ConsensusDocs 301. For example, ¶3.2.9, 4.3.19; AIA §4.8.2.
78 See, for example, VAFR-BIM. 3.3
79 Ashcraft, 178.
80 *Infra* 1.6.
81 ¶3.
82 ¶5.2.
83 ¶1.3.
84 ¶5.7.
85 §3.5.

86 Article 2.
87 §4.
88 528 F.3d 1258 (10th Cir. 2008).
89 Ibid. 1261.
90 Brunka, C. "The Drawing is Mine! The Challenges of Copyright Protection in the Architectural World." *The Journal of Law, Technology & Policy* (2011): 169.
91 Section 1.1.7.
92 AIA Document Commentary A201-2007. 6.
93 AIA B101 §7.
94 §2.3.
95 §2.1.
96 §1.4.8.
97 ¶6.1.
98 ¶6.4.
99 OH-BIM. 9. Reviewed *infra* 3.1.3.4.
100 Ibid. 19.

References

American Institute of Architects. AIA Document Commentary, A201-2007.

American Institute of Architects. A201 (2007) *General Conditions of the Contract for Construction*.

American Institute of Architects. B101 (2007) *Standard Form of Agreement Between Owner and Architect*.

American Institute of Architects. E202 (2008) *Building Information Modeling Protocol Exhibit*.

American Institute of Architects. E203 (2013) *Building Information Modeling and Digital Data Exhibit*.

American Institute of Architects. G201 (2013) *Project Digital Data Protocol Form*.

American Institute of Architects. G202 (2013) *Project Building Information Modeling Protocol Form*.

Ashcraft, H. W. "Building Information Modeling: A Framework for Collaboration." *Construction Law* 28 (2008): 5.

Bernstein, P. G., & Pittman, J. H. "Barriers to the Adoption of Building Information Modeling in the Building Industry." *Autodesk Building Solutions* (2004).

BIMForum. *Level of Development Specification, Draft 1*. April 19 2013.

Brunka, Christina. "The Drawing is Mine! The Challenges of Copyright Protection in the Architectural World." *The Journal of Law, Technology & Policy* (2011): 169.

Build Smart. Singapore Building and Construction Authority. Issue 9. December 2011: 4.

Colorado. Revised Statutes. 12-25-302 (7) (2013).

Industry Foundation Class (IFC). Accessed September 6, 2014. http://www.buildingsmart.org/standards/ifc/model-industry-foundation-classes-ifc.

Larson, D. A., & Golden, K. A. "Entering the Brave, New World: An Introduction to Contracting for Building Information Modeling." *William Mitchell Law Review* 34 (2007): 75.

Lowe, R. H., & Muncey, J. M. "ConsensusDOCS 301 BIM Addendum." *Construction Law* 29 (2009): 17.

Mahoney, J. J. *Barriers to CADD in the ACE Industry*. Stanford. CIFE. Technical Report 23 (1990).

National Council of Architectural Registration Boards. (NCARB) *2014–2015 Rules of Conduct*. Rev. July 2014. Rule 5.2.

National Society of Professional Engineers. *Position Statement No. 1745—Responsible Charge*. Adopted: April 2005; latest revision: July 2010.

Pennsylvania Code. 49 Pa. Code §9.142.

Post, N. "BIM Lawsuit Offers Cautionary Tale." *Architectural Record* (2011).

Post, N. "Help Coming for BIM Users." *Engineering News Record*. May 20, 2013.

Scolari, M. *Oblique Drawing: A History of Anti-perspective*. Cambridge: MIT Press (2012).

Ten Books on Architecture. Trans. Morris H. Morgan, Ph.D, LL.D. Harvard University. (1914). Full text, Project Gutenberg. Accessed on September 6, 2014. http://www.gutenberg.org/ebooks/20239.

Walker, S. G., Holderness, R. A., & Butler, S. D. *State-By-State Guide to Architect, Engineer, and Contractor Licensing*. Aspen Law & Business, 1999.

List of cases

MA Mortenson Co. v. Timberline Software Corporation, 970 P.2d 803, 93 Wash. App. 819 (Ct. App. 1999)

Meshwerks, Inc. v. Toyota Motor Sales USA, Inc., 528 F.3d 1258 (10th Cir. 2008)

RW Vaught Co. v. FD Rich Co., 439 F.2d 895 (Court of Appeals 1971)

Taylor v. DeLosso, 725 A.2d 51, 319 N.J. Super. 174, 319 N.J. (Super. Ct. App. Div. 1999)

Wynner v. Buxton, 97 Cal. App. 3d 166, 158 Cal. Rptr. 587 (Ct. App. 1979)

5 Methods and techniques for analysis of claims involving BIM/VDC

Where claims involving BIM/VDC may exist, the essential framework for analyzing those claims is no different than that of a traditional claim. The same three-legged stool applies: entitlement, causation, damages. A party must demonstrate a legal right to bring a claim in the first instance. They must then prove a causal link between events that are the liability of others and the increased costs. Finally, and significantly, there must be a calculation of the damages. As many have noted, a liability narrative, no matter how compelling, is nothing without proof of damages. As with the literature on construction claims in general, the subset of literature dedicated to damages is equally voluminous. A brief summary overview is presented here for context.

In calculating damages the parties will seek to account for both direct and indirect damages. Direct damages include items such as additional material, equipment, and labor costs. In-direct damages include items such as lost productivity. Whereas documentation supporting direct costs is usually more readily tracked throughout a project and available for review (e.g. invoices, purchase orders, time sheets, etc.) the calculation of lost productivity is highly contentious because it is typically neither tracked, nor easily discerned.[1] For example, when a project is impacted by an event that causes subcontractors from multiple trades to be "stacked," that is, crowded together working in the same physical space, it is difficult to say with exacting precision just what the lost productivity of the electrician was by the fact of his having to wait for the plumber to move out of his way. Likewise, if trades are forced to work excessive overtime in order to keep to schedule as a result of errors or omissions in the design drawings which then leads to a decline in productivity, how does one measure such a loss in productivity? Questions such as these only increase in complexity when there are multiple cumulative impacts across a project that may have any number of causes, including: labor availability, out of sequence work, untimely owner approvals, and so on.[2] As the saying goes, "death by a thousand cuts."

Typical methods for calculating damages and establishing a causal link include the "total cost approach," "modified total cost," "discrete cost," and "quantum meruit."[3] At a broad stroke, the total cost approach calculation determines the excess of the actual costs over the planned costs. Modified total cost is the same as the total cost method, except the contractor subtracts out any acknowledged problems, such as known bid errors, costs for failing to mitigate damages in the field, and so on. A discrete cost method looks at the costs related to a specific event. Quantum meruit considers the work required to have been significantly different from the work as bid, such that the original estimate is no longer applicable.

When calculating lost productivity the method most widely utilized and regularly accepted by the courts is known as the measured mile calculation.

> This calculation compares identical activities in impacted and non-impacted sections of the project in order to ascertain the loss of productivity resulting from the impact of a known set of events. The Measured Mile calculation is favored because it considers only the actual effect of the alleged impact and thereby eliminates disputes over the validity of cost estimates, or factors that may have impacted productivity due to no fault of the owner.[4]

Informed by a baseline appreciation for the entitlement–causation–damages framework and a cursory look at calculating damages, the primary questions in claims analysis might be distilled as:

- What did the *contract* require/provide for?
- What are the applicable *standard of care* and *workmanlike performance*?
- What was the *planned* course of events?
- What was the *actual* course of events?
- What's the *difference* between planned and actual?
- What's the *impact* of the difference
- Who's *responsible* for the impact?
- What's the *dollar* value of the impact?

5.1 Reviewing contract documents for BIM and VDC responsibilities

Contracts must be reviewed to determine BIM/VDC specific duties, responsibilities, and requirements. In addition to any specific exhibits or addenda, the main body of the contract should be reviewed to ensure that any BIM/VDC provisions included therein are identified. Similarly, governing contracts must be reviewed for any "digital" requirements or provisions, including those regarding, the transmission of digital data of any type, data loss, archiving and preservation, and so on. Attention should also be paid to instances where the terms "CAD" or "computer-aided design and drafting (CADD)" or "computer" or derivatives thereof are used in the governing contract. Certain legacy contracts might treat CAD and BIM as equal organizing terms/headers with very little distinction, but which have altered substantive requirements related to BIM/VDC under the headings of "CAD."

Needless to say, all BIM/VDC-specific exhibits, addenda, riders, and so on need to be carefully reviewed. For example, in the case of form documents such as those from the AIA and ConsensusDocs those documents should be compared against published originals to reveal any project-specific alterations, additions, or subtractions that the parties have made. Similarly, those form documents are intended to work in conjunction with additional contract documents – the AIA E203 (2013) with potentially both the G201 (2013) and G202 (2013) protocols; and the ConsensusDocs 301 with a BEP which then becomes an amendment to the *Addendum*.[5] Any amendments to BIM/VDC exhibits and addenda must also be reviewed.

In many instances contractors have developed BIM/VDC riders that are attached to their trade-specific subcontracts. These documents need to be reviewed and understood in terms of how they interact with other BIM/VDC exhibits, addenda, or contract language. Also, contracts held directly by the owner should be reviewed for their BIM/VDC requirements. For example, hospital owners may contract directly with medical equipment consultants who typically have BIM content development requirements. Aside from design and construction considerations, those BIM deliverables could potentially impact post-construction facilities management uses that the parties have also contracted for. Owners might also contract with surveyors who will utilize point cloud scanning technologies to document existing conditions. How BIM/VDC deliverables from owner consultants interact with the BIM/VDC content of both the design and the construction team can be critical.

In addition to contract documents, any pre-contract or pre-bid documents containing BIM/VDC components may also prove insightful and should be reviewed. Examples might include requests for qualifications (RFQs), requests for proposals (RFPs), or pre-bid meeting notes, minutes and/or presentations. Some have questioned if pre-contract communications that create an express warranty that exceeds professional standards implied by law might withstand and supersede lesser or typical standards in the final contract.[6]

5.2 Applicable standards of care and workmanlike performance

Along with contract-specific requirements, analysis will inherently contemplate the BIM/VDC standard of care for design professionals and workmanlike performance for contractors. Typical and customary questions in evaluating performance and conduct in this regard include:

- Have the implied and explicit terms and performance criteria of the engagement been met?
- Are the documents suitable for their intended purposes whether it be pricing, bidding, or construction?
- Are the documents reasonably complete, coordinated, and internally consistent?
- Do the plans and specifications adequately define quantity, configuration, and quality?
- Has adequate interdisciplinary design coordination been effectively and timely implemented?

Chapter 3 of this book focused exclusively on investigating reasonable expectations for standard of care and workmanlike performance as would likely be supported across the broadest geographic spectrum. The analysis included a review of selected primary (form document) and secondary (guidelines, etc.) BIM/VDC source documents over approximately the last 20 years. Collectively and individually these evidentiary sources would appear to support expectations of standard of care and workmanlike performance regarding BIM/VDC that include, at a minimum:

- The use of BIM authoring tools as the primary production tool for producing more consistent and better coordinated 2D contract document sets.
- The use of project review tools to improve systems coordination via 3D clash detection.

Indeed, as previously noted, during the course of writing of this book (June 2014), the National Council of Architectural Registration Boards (NCARB) posted commentary related to decreasing the required apprenticeship hours for their Intern Development Program (IDP). As a direct result of the impact of technology on architectural practice NCARB proposals were considering a 33 percent reduction in the hours for training and preparing architects for licensure.[7]

Given the AIA definition for standard of care which evaluates an architect's performance against architects "practicing in the same or similar locality under the same or similar circumstances" architects might be compared against national best practices as consistent with their locality tranche which, in turn, may have higher BIM/VDC standards.[8]

5.3 Analyzing the planned course of events

All applicable schedules must be reviewed for an understanding of how they include or exclude planned and/or expected BIM/VDC activities. For example, design professional contracts may include typical and customary requirements for regularly scheduled design review meetings. While BIM/VDC tools and processes may be utilized during any such meetings, are there also additional requirements for BIM/VDC specific meetings? For example, the E203 (2013) requires the parties to provide a detailed description of design coordination and clash detection procedures.[9] Likewise, the ConsensusDocs 301 (2008) requires project participants to establish a schedule for the initial delivery and updating of each model within the BEP.[10] Have necessary contractual protocols and processes regarding BIM/VDC-specific schedule requirements been accounted for in overall project schedules? The possibility exists that master schedules might be prepared with only a cursory nod towards BIM/VDC activities. Also, if the parties agreed to utilize 4D BIM/VDC in order to visualize time-sequenced schedules, any applicable schedule milestones and corresponding BIM deliverables should be examined. Likewise, any contemporaneous documentation related to the continuous updating of the schedule and model should be reviewed. As seen in Chapter 4, the courts have ruled on issues involving construction schedules prepared with technology.

5.4 Investigating the actual course of BIM and VDC events

Determining the actual course of events requires a balanced review of all relevant contemporaneous project documentation in light of the contracts and schedules as noted above. And, for all the bells and whistles of digital models, a review of non-model documentation will be vital to appreciating the complete context of any BIM/VDC deliverable. A list of documents would include:

- contract documents – drawings and specifications;
- bid documents;
- project bulletins and addendums;
- BIM content:
  - o   native formats;
  - o   compiled formats for viewing;
- project financials;
- applications for payment;
- requests for information (RFIs) and logs;
- change orders;
- daily reports;
- correspondence;
- notes and meeting minutes;
- access to BIM collaboration solutions/servers/file transfer protocol (FTP) sites, and so on;
- access to electronic document management systems.

A physical review of any and all 2D printed drawings should be done. For example, the standardized graphic composition of 2D plans produced by and extracted from Autodesk® Revit® models are immediately recognizable to most professionals familiar with the use of that tool. It would not be an exaggeration to say that in many instances such a professional thumbing through a set of construction documents could, merely by looking at physical printouts, likely tell at a high level first pass which drawings came from a BIM produced with Autodesk® Revit®, and those drawings that were produced using traditional CAD tools such as AutoCAD®. Reviewing the physical 2D drawings is a useful first step in evaluating BIM content as this can provide, amongst other information, preliminary understanding as to which drawings may have been produced distinct from the BIM process and then inserted into contract sets after the fact. Addenda and bulletin update drawings should be reviewed to assist in determining if any original construction documents produced using BIM had, by the issuance of updates, been abandoned in favor of traditional CAD programs. Marking up the physical drawings with specific areas and issues that are the basis of the claim will also assist in conducting an organized review of the BIM content itself.

The level of organization of a project's BIM content can impact the speed of review of that content. On projects where a BIM/VDC collaboration technology has been utilized the process is typically more streamlined. Depending on the given collaboration solution, both design and construction BIMs might be indexed and searchable at the root level. For example, one can quickly sequester all pitched plumbing pipe, or structural steel models on the fourth floor. Additionally (and significantly for reconstructing timelines of events), each of the BIMs might indicate the individual user responsible for uploads/downloads and be date/time stamped. If a claim includes damages for lost productivity as a result of poorly coordinated contract drawings, increased upload/download of BIM content by given trades might provide corroborating support establishing a causal link between the deficient drawings, the need of subcontractors to continuously coordinate the work at that location within the BIM, and the resulting lost productivity damage calculations. Recall that the AIA E203 (2013) provides parties the opportunity to utilize an electronic document management system.[11] Such a system may or may not support BIM content storage. Likewise where such systems support BIM content storage they may not necessarily provide the type of root level indexing and search capabilities described above. The E203 (2013) in conjunction with the associated G202 (2013) protocol also enable projects utilizing BIM to establish protocols related to model exchange, storage, and so on.[12] The AIA documents do not specifically require a system that provides user/date/time stamping of model uploads/downloads. The ConsensusDocs 301 (2008) does require user/date/time stamping of model collaboration system entries.[13]

Documentation and logs of RFIs and change order documentation must be reviewed and reflected against any contractual BIM/VDC requirements for incorporating changes back into models. For example, the AIA E203 (2013) has general requirements concerning the resolution of changes to a model.[14] The ConsensusDocs 301 (2008) has specific requirements for utilizing BIM in the RFI and change order processes, including response protocols and timing for incorporating necessary changes into any model.[15]

5.5 Comparing planned versus actual events, determining the impact, and explaining the results

A goal of comparing planned versus actual events is determining the root cause of delays, cost overruns or inefficiencies and then isolating the impact of specific events on direct and in-direct costs. In evaluating claims that either focus on BIM/VDC issues *in and of themselves*, or claims that happen to also contain BIM/VDC elements, BIM tools and VDC processes can bring transparency and clarity as a result of their inherent power to visualize. For example, if design documents are alleged to be uncoordinated and inconsistent and there is an associated BIM, the BIM will likely bring some level of initial clarity by being able to visualize the alleged coordination deficiencies. As seen in Chapter 2, the linking of D-BIMs enables the design team to see the scope of other consultants in 3D and provides a clearer picture of overall coordination. The architect's proposed ceiling height "understands" its spatial relationship to steel members. If the MEP drawings are alleged deficient, and there is no corresponding BIM for MEP, issues might be more clearly understood. Or, if a lack of BIM skills of a particular trade contractor are alleged to have caused delay by virtue of being un-prepared for scheduled coordination meetings, or requiring re-work in the field as a result of failing to properly include elements of their scope of work within a BIM for coordination, the applicable models will immediately visualize those deficiencies in ways that 2D drawings typically do not. Furthermore, the ability to visualize the cumulative impact of discrete isolated events across the entire project is greatly enhanced by a 3D model that can instantaneously provide any orthogonal or perspective view in isolation, or within a larger context.

To be sure, charts and diagrams are essential tools in articulating issues to triers of fact, whether they be other project participants during negotiations, or mediators, arbitration panels, or juries. With respect to the admissibility of computer visualizations as evidence during trial, there is precedent. For example, a New York case from 1984, *The People of the State of New York* v. *Michael McHugh*[16] addressed whether a computer re-enactment of a fatal car crash was admissible as evidence in a criminal trial. The court determined that the computer re-enactment was akin to a chart or diagram and reasoned:

> Whether a diagram is hand drawn or mechanically drawn by means of a computer is of no importance . . . A computer is not a gimmick and the court should not be shy about its use when proper. Computers are simply mechanical tools – receiving information and acting on instructions at lightning speed. When the results are useful, they should be accepted, when confusing, they should be rejected.[17]

More recently, the US District Court for the Eastern District of New York in *Garner Allen* v. *Dale Artus, Superintendent*[18] affirmed the trial court's decision to allow a computer animation which incorporated a 3D architectural model generated from 2D floor plans as admissible evidence in a murder trial. For trial an exhibit was prepared by a computer animation teacher to create a timeline sequence of the petitioner's and victim's movements leading up to and after the crime. The computer animation teacher testified to using Autodesk Maya® software to create the 3D model from 2D drawings which was then overlaid with highlights from surveillance cameras. The 2D floor plan used to generate the model was verified through the testimony of a library employee. Citing, amongst others, *People* v. *McHugh* the court reasoned that since the surveillance footage was not edited and drew no conclusions about any of the victim's or petitioner's movements, the animation "helped jurors visualize the relative movements of the victim and petitioner through the library more accurately." It would appear a reasonably short theoretical jump from aiding triers of fact in accurately visualizing the movements of people to assisting parties with visualizing root causes of acceleration, delay, disruption, professional liability, and so on in construction claims. As more construction claims embrace the use of BIM/VDC in the preparation of summary graphics and support documentation, the direction of the *McHugh* court as cited above remains relevant and instructive – "When the results are useful, they should be accepted, when confusing, they should be rejected."[19]

5.6 Calculating BIM/VDC-specific damages

As noted in the overview of disruption claims in Chapter 1, the Mechanical Contractors Association of America (MCAA) *Change Orders, Productivity, Overtime: A Primer for the Construction Industry (2012)*[20] specifically identifies the BIM process as another type of labor that can be impacted by project events and that such costs should be carefully reviewed in the preparation of delay and/or disruption claims. Documentation supporting direct damages for labor related to BIM/VDC will likely be tracked and accessible for review. As with traditional claims, proving lost productivity related to BIM/VDC labor will likely remain highly contentious. Furthermore, the industry will probably need to confront theoretical questions such as whether the standard measured mile method of calculating lost productivity, which is based on actual field-work unit installation, is applicable to virtual unit installation BIM/VDC labor.

5.7 Scenario exploration

The following thought experiments are offered as a means to critical evaluation of the categorization of BIM/VDC legal issues from Chapter 4 and the claims analysis methods above.

Scenario

An architect is on the board of directors for a local BIM/VDC industry organization, an affiliation which is displayed prominently on his firm's website. The architect wins a new hospital project in his home city, a mid-tier market, and has promised the owner, "We'll do BIM." The contract with the owner contains loose language regarding BIM requirements. The architect does not require his MEP consultant to utilize BIM.

The contractor on the project also has a fixed-fee contract for pre-construction services. As part of those pre-construction services the contractor is allowed to review the architect's BIM for reference only.

The project completes on schedule, but the contractor makes a negligent misrepresentation claim against the architect for deficiencies in MEP contract drawings that required a significant design issue to be solved in parallel with coordination and created significant BIM content creation disruptions.

Query

Has the architect met practice standards for BIM/VDC regarding hospital design in a mid-tier market?

How might the deficient MEP drawings be viewed in light of current illustrations in the *Restatement (Second) of Torts*?

Scenario

On a university project for a new student union building the design team of architect and mechanical engineer is highly accomplished in BIM delivery, collectively having completed fifteen BIM/VDC projects together. One of those projects was recognized by the local AIA chapter for having produced high-quality construction documents extracted from a BIM. The structural engineer is not BIM savvy. All parts of the project are well coordinated with exception of the kitchen which is delayed significantly by owner changes and foundation issues which occur underneath the kitchen.

The construction manager on the project has BIM experience, although this is currently the only BIM project currently in the office. The construction manager is acting as the BIM coordinator. The MEP subcontractor has BIM experience on four previous projects. The steel detailer has significant BIM experience. During the course of pre-construction coordination the transition from the design steel model to the detailing model becomes cumbersome as final foundation and kitchen design issues compete for resolution. The MEP subcontractor brings in additional manpower from an overseas vendor to create and coordinate the BIM content needed to complete the kitchen work. The vendor misses deadlines in delivering BIM content and models geometry to a level of detail greater than what is required for coordination (e.g. small diameter conduit, etc.).

The project is significantly delayed and the contractor makes a claim for delay and disruption damages including BIM/VDC labor inefficiencies.

Query

What method of damages calculation might be most applicable in trying to account for lost productivity of the MEP subcontractor?

Scenario

A large historical estate is undergoing significant modernization and refurbishment into a private residence. An architect specializing in this type of work intends to deliver the contract drawings utilizing BIM, although this will be the first BIM project for the office. The owner's representative drafts a BEP which requires the architect to extract all drawings from a BIM. Likewise, there is a clause stating that final payment will be withheld pending the delivery of, amongst other items, a conformed design BIM to incorporate all changes in the contractor's as-built drawings. The BEP is amended to the governing contract.

After 50 percent design development, the architect abandons BIM, and delivers all remaining drawings for all phases via a traditional 2D CAD program. The project completes with minor delay. At the conclusion of the project the architect completes a few basic portions of the BIM as abandoned at 50 percent design development. The architect then imports the entire 2D record drawing set into the BIM and claims that he has met the BEP requirement for all drawings to be extractions from the BIM and a conformed design BIM. He bases his assertion, in part, on the fact that any section drawing extracted from the BIM would provide the substantive dimensional accuracy required for any future improvements, and does not differ significantly from the final construction documents. The owner withholds payment.

Query

Has the architect met his contractual obligations for a conformed design BIM?

Notes

1 AACE International. *Estimating Lost Labor Productivity in Construction Claims.* Recommended Practice No. 25R-03. 2004.
2 Ibid. 5.
3 For example, Ibbs, W., & Nguyen, L. D. "Using the Classical Measured Mile Approach and Variants to Quantify Cumulative Impact Claims." *Construction Law* 32 (2012): 18. Cushman, R. F. *Proving and Pricing Construction Claims.* Aspen Publishers (1996).
4 AACE. *Estimating Lost Labor* 11. Citing Schwartzkopf, W., & McNamara, J. J. *Calculating Construction Damages.* Aspen Publishers Online (2001).
5 *Supra* 3.1.4.2, 3.1.4.5.
6 Sweet, J. *Sweet on Construction Industry Contracts: Major AIA Documents.* Wiley (1996): 168.
7 http://blog.ncarb.org/2014/June/IDP-Proposals.aspx. Accessed on December 14, 2014.
8 AIA B201 (2007) §2.2.
9 AIA G202 (2013) §1.7.7.
10 ConsensusDocs 301 (2008) ¶4.3.6, 4.3.7.
11 §3.5.1.
12 E203 (2013) Article 4, generally.
13 ConsensusDocs 301 §3.2.5.
14 E203 (2013) §4.5.1.7.
15 ConsensusDocs 301 (2008) ¶4.3.21, 4.3.22, respectively.
16 124 Misc. 2d 559, 476 N.Y.S.2d 721, 476 N.Y.S. 721 (NY: Supreme Court, Bronx 1984).
17 Ibid. 560.
18 No. 09-CV-4562 (JFB) (E.D.N.Y. May 14, 2014).
19 *McHugh.* 560.
20 *Management Methods Bulletin C02-2011.* MCAA. (2012).

References

AACE International. *Estimating Lost Labor Productivity in Construction Claims.* Recommended Practice No. 25R-03 (2004).

American Institute of Architects. E203 (2013) *Building Information Modeling and Digital Data Exhibit.*

American Institute of Architects. G202 (2013) *Project Building Information Modeling Protocol Form.*

ConsensusDocs. 301 *Building Information Modeling (BIM) Addendum* (2008).

Ibbs, W., & Nguyen, L. D. "Using the Classical Measured Mile Approach and Variants to Quantify Cumulative Impact Claims." *Construction Law* 32 (2012).

Mechanical Contractors Association of America. *Management Methods Bulletin C02-2011.* MCAA (2012).

Sweet, Justin. *Sweet on Construction Industry Contracts: Major AIA Documents.* Wiley (1996).

List of cases

Garner Allen v. *Dale Artus, Superintendent*, No. 09-CV-4562 (JFB) (E.D.N.Y. May 14, 2014)

The People of the State of New York v. *Michael McHugh*, 124 Misc. 2d 559, 476 N.Y.S.2d 721, 476 N.Y.S. 721 (NY: Supreme Court, Bronx 1984)

6 Preventative measures for enabling BIM/VDC success

Any narrative that accompanies a discussion regarding successful implementation of BIM/VDC invariably includes "communication" as a key component. BIM/VDC, as the saying goes, "Is 10 percent technology, 90 percent psychology." Accordingly, below are a series of pragmatic checklists. These questions are meant to be additional prompts to aid parties in clarifying BIM/VDC expectations, duties, and requirements.

6.1 Owners and facility managers

☐ Will there be specific BIM/VDC requirements for both design and construction?
☐ Have the BIM/VDC qualifications of all parties been fully vetted and verified?
☐ Is counsel involved with the drafting of any requests for proposal (RFPs), contracts, and so on familiar with BIM/VDC?
☐ If boiler-plate BIM/VDC documents will be utilized, what were the selection criteria?
☐ Do corresponding sections of governing contracts align with BIM/VDC requirements?
☐ What is the status of a BIM in regards to contract documents?
☐ How specifically will BIM support the change order process?
☐ What mechanisms are in place to monitor contract administration in respect of BIM/VDC?
☐ Are internal staff prepared to support the project team's BIM/VDC needs?
☐ With respect to BIM for facilities, what *exactly* are the requirements of the owner?
☐ Are compensation expectations with respect to BIM/VDC set?
☐ Do BIM/VDC requirements align with the current design standards of the owner?

6.2 Architects and engineers

☐ Will staffing allow BIM/VDC personnel to remain on a given project for the duration?
☐ Are consultants fully BIM/VDC capable with verifiable experience?
☐ Are executives associated with the job BIM/VDC educated?
☐ Are the project architect and/or project manager committed to BIM/VDC success?

- ☐ Is there IT infrastructure in place to support the needs of the project?
- ☐ Is counsel involved with the drafting of any RFPs, contracts, and so on familiar with BIM/VDC?
- ☐ If boiler-plate BIM/VDC documents will be utilized, what were the selection criteria?
- ☐ Do corresponding sections of governing contracts align with BIM/VDC requirements?
- ☐ Are compensation expectations fully articulated?
- ☐ What is the status of a BIM in regards to contract documents?
- ☐ Are the requirements for conformed design BIMs?
- ☐ Have the BIM/VDC requirements of the owner's consultants been shared?
- ☐ Are there data sharing and licensing provisions in place?
- ☐ Is there a requirement for a BIM Execution Plan?

6.3 Contractors and general contractors

- ☐ Have the BIM/VDC qualifications of all subcontractors been fully vetted and verified?
- ☐ Is counsel involved with the drafting of any contracts, riders, and so on familiar with BIM/VDC?
- ☐ Are executives associated with the job BIM/VDC educated?
- ☐ If boiler-plate BIM/VDC documents will be utilized, what were the selection criteria?
- ☐ Do corresponding sections of governing contracts align with BIM/VDC requirements?
- ☐ What is the status of a BIM in regards to contract documents?
- ☐ How specifically will BIM support the request for information process?
- ☐ How specifically will BIM support the change order process?
- ☐ Will mobile solutions be used for field execution of work and are those systems tied to BIM?
- ☐ What mechanisms are in place to monitor contract administration in respect of BIM/VDC?
- ☐ Are internal staff prepared to support the project team's BIM/VDC needs?
- ☐ With respect to BIM for facilities, what *exactly* are the owner's requirements?
- ☐ Are compensation expectations with respect to BIM/VDC set?
- ☐ Do BIM/VDC requirements align with the current design standards of the owner?
- ☐ How will estimators utilize the BIM?

6.4 Subcontractors and fabricators

- ☐ Have the BIM/VDC qualifications of any third-party consultants been vetted and verified?
- ☐ Is counsel involved with the drafting of any contracts, riders, and so on familiar with BIM/VDC?
- ☐ Are executives associated with the job BIM/VDC educated?
- ☐ Have the specific requirements of any riders been evaluated?

☐ If boiler-plate BIM/VDC documents will be utilized, what were the selection criteria?

☐ Do corresponding sections of governing contracts align with BIM/VDC requirements?

☐ What is the status of a BIM in regards to contract documents?

☐ How specifically will BIM support the request for information process?

☐ How specifically will BIM support the change order process?

☐ Will mobile solutions be used for field execution of work and are those systems tied to BIM?

☐ What mechanisms are in place to monitor contract administration in respect of BIM/VDC?

☐ With respect to BIM for facilities, what *exactly* are the owner's requirements?

☐ Are compensation expectations with respect to BIM/VDC set?

☐ What is the legal status of the design team's BIM content?

6.5 Drivers of change and impacts to claims involving BIM up to 2025

This book began with the assertion that design and construction are "risky." This will undoubtedly remain the case regardless of what technology or processes are employed. However, BIM/VDC have proven themselves effective at alleviating risk. Cumulatively, the primary and secondary source documents from across the past two decades appear to reasonably support baseline BIM/VDC expectations of standard of care and workmanlike performance. When properly implemented alongside general best practices for design and construction, BIM/VDC has shown an ability to produce better coordinated drawings sets with fewer errors and omissions, thereby enabling more efficient pre-construction coordination which, in turn, leads to fewer requests for information and change orders. Cumulatively, the end result enables stakeholders to better control cost and schedule.

Form documents from both the AIA and ConsensusDocs continue to be updated, signaling that the AEC market is implementing BIM/VDC at the contract level. And while form documents and publically available BIM/VDC guidelines will not be enough to complete a transformation of the industry, shifts in the professional licensing structure for architects may be initiating as well. As noted several times above, NCARB has publically considered a proposal to reduce the number of internship hours for architects as a direct result of technology's impact on practice. Similarly, the contracting community has seen value in offering BIM certifications, such as the AGC's BIM certificate program. Along with these initiatives, organizations such as FIATECH pursue common processes for e-submittals of BIM for code review and compliance. In light of the above it can only be a matter of time before the insurance and surety markets develop BIM/VDC-specific instruments. With the pace of development for BIM/VDC business processes and systems continuing to move in concert with the development of the technologies themselves, the next decade will surely see continued advancements for the AEC industry that impact claims and disputes.

Appendix A

AIA® Document E202™ – 2008

Building Information Modeling Protocol Exhibit

This Exhibit is incorporated into the accompanying agreement (the "Agreement") dated the day of in the year
(In words, indicate day, month and year.)

BETWEEN:
(Name, address and contact information, including electronic addresses)

This document has important legal consequences. Consultation with an attorney is encouraged with respect to its completion or modification.

AND:
(Name, address and contact information, including electronic addresses)

for the following Project:
(Name and location or address)

TABLE OF ARTICLES

1 GENERAL PROVISIONS

2 PROTOCOL

3 LEVEL OF DEVELOPMENT

4 MODEL ELEMENTS

ARTICLE 1 GENERAL PROVISIONS

§ 1.1 This Exhibit establishes the protocols, expected levels of development, and authorized uses of Building Information Models on this Project and assigns specific responsibility for the development of each Model Element to a defined Level of Development at each Project phase. Where a provision in this Exhibit conflicts with a provision in the Agreement into which this Exhibit is incorporated, the provision in this Exhibit will prevail.

§ 1.1.1 The parties agree to incorporate this Exhibit by reference into any other agreement for services or construction for the Project.

§ 1.2 Definitions
§ 1.2.1 Building Information Model. A Building Information Model(s) is a digital representation of the physical and functional characteristics of the Project and is referred to in this Exhibit as the "Model(s)," which term may be used herein to describe a Model Element, a single Model or multiple Models used in the aggregate. "Building Information Modeling" means the process and technology used to create the Model.

§ 1.2.2 Level of Development. The Level(s) of Development (LOD) describes the level of completeness to which a Model Element is developed.

§ 1.2.3 Model Element. A Model Element is a portion of the Building Information Model representing a component, system or assembly within a building or building site. For the purposes of this Exhibit, Model Elements are represented by the Construction Specifications Institute (CSI) UniFormat™ classification system in the Model Element Table at Section 4.3.

§ 1.2.4 Model Element Author. The Model Element Author is the party responsible for developing the content of a specific Model Element to the LOD required for a particular phase of the Project. Model Element Authors are identified in the Model Element Table at Section 4.3.

§ 1.2.5 Model User. The Model User refers to any individual or entity authorized to use the Model on the Project, such as for analysis, estimating or scheduling.

ARTICLE 2 PROTOCOL
§ 2.1 Coordination and Conflicts
Where conflicts are found in the Model, regardless of the phase of the Project or LOD, the discovering party shall promptly notify the Model Element Author(s). Upon such notification, the Model Element Author(s) shall act promptly to mitigate the conflict.

§ 2.2 Model Ownership
In contributing content to the Model, the Model Element Author does not convey any ownership right in the content provided or in the software used to generate the content. Unless otherwise granted in a separate license, any subsequent Model Element Author's and Model User's right to use, modify, or further transmit the Model is specifically limited to the design and construction of the Project,

and nothing contained in this Exhibit conveys any other right to use the Model for another purpose.

§ 2.3 Model Requirements

§ 2.3.1 Model Standard. The Model shall be developed in accordance with the following standard, if any: (*Set forth below object naming conventions, graphic standards, common symbology, etc., or state an applicable standard, such as the National Building Information Model Standards (NBIMS).*)

§ 2.3.2 File Format(s). Models shall be delivered in the following format(s) as appropriate to the use of the Model:

Use of Model	Required File Format(s)

§ 2.4 Model Management

§ 2.4.1 The requirements for managing the Model include, but are not limited to, the duties set forth below in this Section 2.4. The Architect will manage the Model from the inception of the Project. If the responsibility for Model management will be assigned to another party at a particular phase of the Project, indicate below the identity of the party that will assume that responsibility, and the phase at which that party will assume those responsibilities.

Responsible Party	Project Phase

§ 2.4.2 Initial Responsibilities. The party responsible for managing the Model shall facilitate the establishment of protocols for the following:

.1   Model origin, coordinate system, and units
.2   File storage location(s)
.3   Processes for transferring and accessing Model files
.4   Clash detection

.5   Access rights
.6   Other protocols:
(*Insert additional protocols below.*)

§ 2.4.3 Ongoing Responsibilities. The party responsible for managing the Model shall have the following ongoing responsibilities:

.1   Collect incoming Models:

.1   Coordinate submission and exchange of Models
.2   Log incoming Models
.3   Validate that files are complete and usable and in compliance with applicable protocols
.4   Maintain record copy of each file received

.2   Aggregate Model files and make available for viewing
.3   Perform clash detection in accordance with established protocols and issue periodic clash detection reports
.4   Maintain Model archives and backups
.5   Manage access rights
.6   Follow protocols established in Section 2.4.2

§ 2.4.4 Model Archives. The party responsible for Model management as set forth in this Section 2.4 shall produce a Model Archive at the end of each Project phase and shall preserve the Model Archive as a record that may not be altered for any reason.

§ 2.4.4.1 The Model Archive shall consist of two sets of files. The first set shall be a collection of individual Models as received from the Model Element Author(s). The second set of files shall consist of the aggregate of those individual Models in a format suitable for archiving and viewing. The second set shall be saved in the following file format:

§ 2.4.4.2 Additional Model Archive requirements, if any, are as follows:

§ 2.4.4.3 The procedures for storing and preserving the Model upon final completion of the Project are as follows:

§ 2.4.5 Other requirements for Model management, if any, are as follows:
(Describe in detail any other Model management requirements.)

ARTICLE 3 LEVEL OF DEVELOPMENT

§ 3.1 The following LOD descriptions identify the specific content requirements and associated authorized uses for each Model Element at five progressively detailed levels of completeness. Each subsequent LOD builds on the previous level and includes all the characteristics of previous levels. The parties shall utilize the five LOD described below in completing the Model Element Table at Section 4.3, which establishes the required LOD for each Model Element at each phase of the Project.

§ 3.2 LOD 100
§ 3.2.1 Model Content Requirements. Overall building massing indicative of area, height, location, and orientation may be modeled in three dimensions or represented by other data.

§ 3.2.2 Authorized Uses
§ 3.2.2.1 Analysis. The Model may be analyzed based on volume, area and orientation by application of generalized performance criteria assigned to the representative Model Elements.

§ 3.2.2.2 Cost Estimating. The Model may be used to develop a cost estimate based on current area, volume or similar conceptual estimating techniques (e.g., square feet of floor area, condominium unit, hospital bed, etc.).

§ 3.2.2.3 Schedule. The Model may be used for project phasing and overall duration.

§ 3.2.2.4 Other Authorized Uses. Additional authorized uses of the Model developed to a Level 100, if any, are as follows:

§ 3.3 LOD 200

§ 3.3.1 Model Content Requirements. Model Elements are modeled as generalized systems or assemblies with approximate quantities, size, shape, location, and orientation. Non-geometric information may also be attached to Model Elements.

§ 3.3.2 Authorized Uses

§ 3.3.2.1 Analysis. The Model may be analyzed for performance of selected systems by application of generalized performance criteria assigned to the representative Model Elements.

§ 3.3.2.2 Cost Estimating. The Model may be used to develop cost estimates based on the approximate data provided and conceptual estimating techniques (e.g., volume and quantity of elements or type of system selected).

§ 3.3.2.3 Schedule. The Model may be used to show ordered, time-scaled appearance of major elements and systems.

§ 3.3.2.4 Other Authorized Uses. Additional authorized uses of the Model developed to a Level 200, if any, are as follows:

§ 3.4 LOD 300

§ 3.4.1 Model Content Requirements. Model Elements are modeled as specific assemblies accurate in terms of quantity, size, shape, location, and orientation. Non-geometric information may also be attached to Model Elements.

§ 3.4.2 Authorized Uses

§ 3.4.2.1 Construction. Suitable for the generation of traditional construction documents and shop drawings.

§ 3.4.2.2 Analysis. The Model may be analyzed for performance of selected systems by application of specific performance criteria assigned to the representative Model Elements.

§ 3.4.2.3 Cost Estimating. The Model may be used to develop cost estimates based on the specific data provided and conceptual estimating techniques.

§ 3.4.2.4 Schedule. The Model may be used to show ordered, time-scaled appearance of detailed elements and systems.

§ 3.4.2.5 Other Authorized Uses. Additional authorized uses of the Model developed to a Level 300, if any, are as follows:

§ 3.5 LOD 400

§ 3.5.1 Model Content Requirements. Model Elements are modeled as specific assemblies that are accurate in terms of size, shape, location, quantity, and orientation with complete fabrication, assembly, and detailing information. Non-geometric information may also be attached to Model Elements.

§ 3.5.2 Authorized Uses

§ 3.5.2.1 Construction. Model Elements are virtual representations of the proposed element and are suitable for construction.

§ 3.5.2.2 Analysis. The Model may be analyzed for performance of approved selected systems based on specific Model Elements.

§ 3.5.2.3 Cost Estimating. Costs are based on the actual cost of specific elements at buyout.

§ 3.5.2.4 Schedule. The Model may be used to show ordered, time-scaled appearance of detailed specific elements and systems including construction means and methods.

§ 3.5.2.5 Other Authorized Uses. Additional authorized uses of the Model developed to a Level 400, if any, are as follows:

§ 3.6 LOD 500

§ 3.6.1 Model Content Requirements. Model Elements are modeled as constructed assemblies actual and accurate in terms of size, shape, location, quantity, and orientation. Non-geometric information may also be attached to modeled elements.

§ 3.6.2 Authorized Uses
§ 3.6.2.1 General Usage. The Model may be utilized for maintaining, altering, and adding to the Project, but only to the extent consistent with any licenses granted in the Agreement or in a separate licensing agreement.

---

§ 3.6.2.2 Other Authorized Uses. Additional authorized uses of the Model developed to a Level 500, if any, are as follows:

ARTICLE 4 MODEL ELEMENTS
§ 4.1 Reliance on Model Elements
§ 4.1.1 The Model Element Table at Section 4.3 identifies (1) the LOD required for each Model Element at the end of each Project phase, and (2) the Model Element Author responsible for developing the Model Element to the LOD identified. Each Model Element Author's content is intended to be shared with subsequent Model Element Authors and Model Users throughout the course of the Project.

§ 4.1.2 It is understood that while the content of a specific Model Element may include data that exceeds the required LOD identified in Section 4.3 for a particular phase, Model Users and subsequent Model Element Authors may rely on the accuracy and completeness of a Model Element consistent only with the content required for the LOD identified in Section 4.3.

§ 4.1.3 Any use of, or reliance on, a Model Element inconsistent with the LOD indicated in Section 4.3 by subsequent Model Element Authors or Model Users shall be at their sole risk and without liability to the Model Element Author. To the fullest extent permitted by law, subsequent Model Element Authors and Model Users shall indemnify and defend the Model Element Author from and against all claims arising from or related to the subsequent Model Element Author's or Model User's modification to, or unauthorized use of, the Model Element Author's content.

§ 4.2 Table Instructions
§ 4.2.1 The table in Section 4.3 indicates the LOD to which each Model Element Author (MEA) is required to develop the content of the Model Element at the conclusion of each phase of the Project.

§ 4.2.2 Abbreviations for each MEA to be used in the Model Element Table are as follows: (*Provide abbreviations such as "A – Architect," or "C – Contractor."*)

§ 4.3 Model Element Table

Identify (1) the LOD required for each Model Element at the end of each phase, and (2) the Model Element Author (MEA) responsible for developing the Model Element to the LOD identified.

Insert abbreviations for each MEA identified in the table below, such as "A – Architect," or "C – Contractor."

NOTE: *LODs must be adapted for the unique characteristics of each Project.*

Model Elements Utilizing CSI UniFormat™				LOD	MEA	LOD	MEA	LOD	MEA	LOD	MEA	LOD	MEA	LOD	MEA	LOD	MEA	Note Number (See 4.4)
A SUBSTRUCTURE	A10 Foundations	A1010	Standard Foundations															
		A1020	Special Foundations															
		A1030	Slab on Grade															
	A20 Basement Construction	A2010	Basement Excavation															
		A2020	Basement Walls															
B SHELL	B10 Superstructure	B1010	Floor Construction															
		B1020	Roof Construction															

§ 4.3 Model Element Table

Identify (1) the LOD required for each Model Element at the end of each phase, and (2) the Model Element Author (MEA) responsible for developing the Model Element to the LOD identified.

Insert abbreviations for each MEA identified in the table below, such as "A – Architect," or "C – Contractor."

NOTE: LODs must be adapted for the unique characteristics of each Project.

Model Elements Utilizing CSI UniFormat™			LOD	MEA	LOD	MEA	LOD	MEA	LOD	MEA	LOD	MEA	LOD	MEA	Note Number (See 4.4)
B20 Exterior Enclosure	B2010	Exterior Walls													
	B2020	Exterior Windows													
	B2030	Exterior Doors													
B30 Roofing	B3010	Roof Coverings													
	B3020	Roof Openings													
C INTERIORS C10 Interior Construction	C1010	Partitions													
	C1020	Interior Doors													

§ 4.3 Model Element Table

Identify (1) the LOD required for each Model Element at the end of each phase, and (2) the Model Element Author (MEA) responsible for developing the Model Element to the LOD identified.

Insert abbreviations for each MEA identified in the table below, such as "A – Architect," or "C – Contractor."

NOTE: LODs must be adapted for the unique characteristics of each Project.

Model Elements Utilizing CSI UniFormat™			LOD	MEA	LOD	MEA	LOD	MEA	LOD	MEA	LOD	MEA	LOD	MEA	Note Number (See 4.4)
	D1020	Escalators & Moving Walks													
	D1030	Other Conveying Systems													
D20	Plumbing	D2010	Plumbing Fixtures													
		D2020	Domestic Water Distribution													
		D2030	Sanitary Waste													
		D2040	Rain Water Drainage													
		D2090	Other Plumbing Systems													

§ 4.3 Model Element Table

Identify (1) the LOD required for each Model Element at the end of each phase, and (2) the Model Element Author (MEA) responsible for developing the Model Element to the LOD identified.

Insert abbreviations for each MEA identified in the table below, such as "A – Architect," or "C – Contractor."

NOTE: LODs must be adapted for the unique characteristics of each Project.

Model Elements Utilizing CSI UniFormat™		LOD	MEA	LOD	MEA	LOD	MEA	LOD	MEA	LOD	MEA	LOD	MEA	Note Number (See 4.4)
D30 HVAC	D3010 Energy Supply													
	D3020 Heat Generating Systems													
	D3030 Cooling Generating Systems													
	D3040 Distribution Systems													
	D3050 Terminal & Package Units													
	D3060 Controls & Instrumentation													
	D3070 Systems Testing & Balancing													

§ 4.3 **Model Element Table**

Identify (1) the LOD required for each Model Element at the end of each phase, and (2) the Model Element Author (MEA) responsible for developing the Model Element to the LOD identified.

Insert abbreviations for each MEA identified in the table below, such as "A – Architect," or "C – Contractor."

NOTE : *LODs must be adapted for the unique characteristics of each Project.*

Model Elements Utilizing CSI UniFormat™			LOD	MEA	LOD	MEA	LOD	MEA	LOD	MEA	LOD	MEA	LOD	MEA	Note Number (See 4.4)
	D3090	Other HVAC Systems & Equipment													
D40	Fire Protection														
	D4010	Sprinklers													
	D4020	Standpipes													
	D4030	Fire Protection Specialties													
	D4090	Other Fire Protection Systems													
D50	Electrical														
	D5010	Electrical Service & Distribution													
	D5020	Lighting and Branch Wiring													

§ 4.3 Model Element Table

Identify (1) the LOD required for each Model Element at the end of each phase, and (2) the Model Element Author (MEA) responsible for developing the Model Element to the LOD identified.

Insert abbreviations for each MEA identified in the table below, such as "A – Architect," or "C – Contractor."

NOTE: LODs must be adapted for the unique characteristics of each Project.

Model Elements Utilizing CSI UniFormat™			LOD	MEA	LOD	MEA	LOD	MEA	LOD	MEA	LOD	MEA	LOD	MEA	Note Number (See 4.4)
		D5030 Communications & Security													
		D5090 Other Electrical Systems													
E EQUIPMENT & FURNISHINGS	E10 Equipment	E1010 Commercial Equipment													
		E1020 Institutional Equipment													
		E1030 Vehicular Equipment													
		E1090 Other Equipment													
	E20 Furnishings	E2010 Fixed Furnishings													

§ 4.3 Model Element Table

Identify (1) the LOD required for each Model Element at the end of each phase, and (2) the Model Element Author (MEA) responsible for developing the Model Element to the LOD identified.

Insert abbreviations for each MEA identified in the table below, such as "A – Architect," or "C – Contractor."

NOTE: LODs must be adapted for the unique characteristics of each Project.

Model Elements Utilizing CSI UniFormat™				LOD	MEA	LOD	MEA	LOD	MEA	LOD	MEA	LOD	MEA	LOD	MEA	Note Number (See 4.4)
			E2020	Movable Furnishings													
F	SPECIAL CONSTR. & DEMO	F10	Special Construction	F1010	Special Structures												
				F1020	Integrated Construction												
				F1030	Special Construction Systems												
				F1040	Special Facilities												
				F1050	Special Controls & Instrumentation												
		F20	Selective Bldg Demo	F2010	Building Elements Demolition												

§ 4.3 Model Element Table

Identify (1) the LOD required for each Model Element at the end of each phase, and (2) the Model Element Author (MEA) responsible for developing the Model Element to the LOD identified.

Insert abbreviations for each MEA identified in the table below, such as "A – Architect," or "C – Contractor."

NOTE: *LODs must be adapted for the unique characteristics of each Project.*

Model Elements Utilizing CSI UniFormat™			LOD	MEA	LOD	MEA	LOD	MEA	LOD	MEA	LOD	MEA	LOD	MEA	Note Number (See 4.4)
		F2020 Hazardous Components Abatement													
G BUILDING SITEWORK	G10 Site Preparation	G1010 Site Clearing													
		G1020 Site Demolition & Relocations													
		G1030 Site Earthwork													
		G1040 Hazardous Waste Remediation													

§ 4.3 Model Element Table

Identify (1) the LOD required for each Model Element at the end of each phase, and (2) the Model Element Author (MEA) responsible for developing the Model Element to the LOD identified.

Insert abbreviations for each MEA identified in the table below, such as "A – Architect," or "C – Contractor."

NOTE: LODs must be adapted for the unique characteristics of each Project.

Model Elements Utilizing CSI UniFormat™			LOD	MEA	LOD	MEA	LOD	MEA	LOD	MEA	LOD	MEA	LOD	MEA	LOD	MEA	Note Number (See 4.4)
G20	Site Improvements																
	G2010	Roadways															
	G2020	Parking Lots															
	G2030	Pedestrian Paving															
	G2040	Site Development															
	G2050	Landscaping															
G30	Site Civil/ Mech. Utilities																
	G3010	Water Supply & Distribution Systems															
	G3020	Sanitary Sewer Systems															
	G3030	Storm Sewer Systems															

§ 4.3 Model Element Table

Identify (1) the LOD required for each Model Element at the end of each phase, and (2) the Model Element Author (MEA) responsible for developing the Model Element to the LOD identified.

Insert abbreviations for each MEA identified in the table below, such as "A – Architect," or "C – Contractor."

NOTE: LODs must be adapted for the unique characteristics of each Project.

Model Elements Utilizing CSI UniFormat™			LOD	MEA	LOD	MEA	LOD	MEA	LOD	MEA	LOD	MEA	LOD	MEA	Note Number (See 4.4)
	G3040	Heating Distribution													
	G3050	Cooling Distribution													
	G3060	Fuel Distribution													
	G3090	Other Civil/ Mechanical Utilities													
G40 Site Electrical Utilities	G4010	Electrical Distribution													
	G4020	Site Lighting													
	G4030	Site Communications & Security													

§ 4.3 Model Element Table

Identify (1) the LOD required for each Model Element at the end of each phase, and (2) the Model Element Author (MEA) responsible for developing the Model Element to the LOD identified.

Insert abbreviations for each MEA identified in the table below, such as "A – Architect," or "C – Contractor."

NOTE: LODs must be adapted for the unique characteristics of each Project.

Model Elements Utilizing CSI UniFormat™			LOD	MEA	LOD	MEA	LOD	MEA	LOD	MEA	LOD	MEA	LOD	MEA	Note Number (See 4.4)
		G4090	Other Electrical Utilities													
	G50	Other Site Construction	G5010	Service Tunnels												
		G5090	Other Site Systems & Equipment													

§ 4.4 Model Element Table Notes

Notes:

(List by number shown on table.)

AIA® Document E203™ – 2013

Building Information Modeling and Digital Data Exhibit

This Exhibit dated the _____ day of _____ in the year _____ is incorporated into the agreement (the "Agreement") between the Parties for the following Project:
(*Name and location or address of the Project*)

This document has important legal consequences. Consultation with an attorney is encouraged with respect to its completion or modification.

TABLE OF ARTICLES

1 GENERAL PROVISIONS

2 TRANSMISSION AND OWNERSHIP OF DIGITAL DATA

3 DIGITAL DATA PROTOCOLS

4 BUILDING INFORMATION MODELING PROTOCOLS

5 SPECIAL TERMS AND CONDITIONS

This document is intended to be incorporated into an agreement between the Parties and used in conjunction with AIA Documents G201™ –2013, Project Digital Data Protocol Form, and G202™–2013, Project Building Information Modeling Protocol Form. It is anticipated that other Project Participants will incorporate a project specific E203–2013 into their agreements, and that the Parties and other Project Participants will set forth the agreed-upon protocols in AIA Documents G201–2013 and G202–2013.

ARTICLE 1 GENERAL PROVISIONS

§ 1.1 This Exhibit provides for the establishment of protocols for the development, use, transmission, and exchange of Digital Data for the Project. If Building Information Modeling will be utilized, this Exhibit also provides for the establishment of the protocols necessary to implement the use of Building Information Modeling on the Project, including protocols that establish the expected Level of Development for Model Elements at various milestones of the Project, and the associated Authorized Uses of the Building Information Models.

§ 1.2 The Parties agree to incorporate this Exhibit into their agreements with any other Project Participants that may develop or make use of Digital Data on the Project. Prior to transmitting or allowing access to Digital Data, a Party may require any Project Participant to provide reasonable evidence that it has incorporated this Exhibit into its agreement for the Project, and agreed to the most recent Project specific versions of AIA Document G201™–2013, Project

Digital Data Protocol Form and AIA Document G202™–2013, Project Building Information Modeling Protocol Form.

§ 1.2.1 The Parties agree that each of the Project Participants utilizing Digital Data on the Project is an intended third party beneficiary of the Section 1.2 obligation to incorporate this Exhibit into agreements with other Project Participants, and any rights and defenses associated with the enforcement of that obligation. This Exhibit does not create any third-party beneficiary rights other than those expressly identified in this Section 1.2.1.

§ 1.3 Adjustments to the Agreement

§ 1.3.1 If a Party believes that protocols established pursuant to Sections 3.2 or 4.5, and memorialized in AIA Documents G201–2013 and G202–2013, will result in a change in the Party's scope of work or services warranting an adjustment in compensation, contract sum, schedule or contract time, the Party shall notify the other Party. Failure to provide notice as required in this Section 1.3 shall result in a Party's waiver of any claims for adjustments in compensation, contract sum, schedule or contract time as a result of the established protocols.

§ 1.3.2 Upon such notice, the Parties shall discuss and negotiate revisions to the protocols or discuss and negotiate any adjustments in compensation, contract sum, schedule or contract time in accordance with the terms of the Agreement.

§ 1.3.3 Notice required under this Section 1.3 shall be provided within thirty days of receipt of the protocols, unless otherwise indicated below:
(If the Parties require a notice period other than thirty days of receipt of the protocols, indicate the notice period below.)

§ 1.4 Definitions

§ 1.4.1 Building Information Model. A Building Information Model is a digital representation of the Project, or a portion of the Project, and is referred to in this Exhibit as the "Model," which term may be used herein to describe a Model Element, a single model or multiple models used in the aggregate, as well as other data sets identified in AIA Document G202–2013, Project Building Information Modeling Protocol Form.

§ 1.4.2 Building Information Modeling. Building Information Modeling or Modeling means the process used to create the Model.

§ 1.4.3 Model Element. A Model Element is a portion of the Model representing a component, system or assembly within a building or building site.

§ 1.4.4 Level of Development. The Level of Development (LOD) describes the minimum dimensional, spatial, quantitative, qualitative, and other data included in a Model Element to support the Authorized Uses associated with such LOD.

§ 1.4.5 Authorized Uses. The term 'Authorized Uses" refers to the permitted uses of Digital Data authorized in the Digital Data and/or Building Information Modeling protocols established pursuant to the terms of this Exhibit.

§ 1.4.6 Model Element Author. The Model Element Author is the entity (or individual) responsible for managing and coordinating the development of a specific Model Element to the LOD required for an identified Project milestone, regardless of who is responsible for providing the content in the Model Element. Model Element Authors are to be identified in Section 3.3, Model Element Table, of AIA Document G202–2013.

§ 1.4.7 Digital Data. Digital Data is information, including communications, drawings, specifications and designs, created or stored for the Project in digital form. Unless otherwise stated, the term Digital Data includes the Model.

§ 1.4.8 Confidential Digital Data. Confidential Digital Data is Digital Data containing confidential or business proprietary information that the transmitting party designates and clearly marks as "confidential."

§ 1.4.9 Written or In Writing. In addition to any definition in the Agreement to which this Exhibit is attached, for purposes of this Exhibit and the Agreement, "written" or "in writing" shall mean any communication prepared and sent using a transmission method set forth in this Exhibit, or the protocols developed pursuant to this Exhibit, that permits the recipient to print the communication.

§ 1.4.10 Written Notice. In addition to any terms in the Agreement to which this Exhibit is attached, for purposes of this Exhibit and the Agreement, "written notice" shall be deemed to have been duly served if transmitted electronically to an address provided in this Exhibit or the Agreement using a transmission method set forth in this Exhibit that permits the recipient to print the communication.

§ 1.4.11 Party and Parties. The terms "Party" and "Parties" refer to the signing parties to the Agreement.

§ 1.4.12 Project Participant. A Project Participant is an entity (or individual) providing services, work, equipment or materials on the Project and includes the Parties.

ARTICLE 2 TRANSMISSION AND OWNERSHIP OF DIGITAL DATA

§ 2.1 The transmission of Digital Data constitutes a warranty by the Party transmitting Digital Data to the Party receiving Digital Data that the transmitting Party is the copyright owner of the Digital Data, or otherwise has permission to transmit the Digital Data for its use on the Project in accordance with the Authorized Uses of Digital Data established pursuant to the terms of this Exhibit.

§ 2.2 If a Party transmits Confidential Digital Data, the transmission of such Confidential Digital Data constitutes a warranty to the Party receiving such Confidential Digital Data that the transmitting Party is authorized to transmit the Confidential Digital Data. If a Party receives Confidential Digital Data, the receiving Party shall keep the Confidential Digital Data strictly confidential and shall not disclose it to any other person or entity except as set forth in Section 2.2.1.

§ 2.2.1 The receiving Party may disclose Confidential Digital Data as required by law or court order, including a subpoena or other form of compulsory legal process issued by a court or governmental entity. The receiving Party may also disclose the Confidential Digital Data to its employees, consultants or contractors in order to perform services or work solely and exclusively for the Project, provided those employees, consultants and contractors are subject to the restrictions on the disclosure and use of Confidential Digital Data as set forth in this Exhibit.

§ 2.3 By transmitting Digital Data, the transmitting Party does not convey any ownership right in the Digital Data or in the software used to generate the Digital Data. Unless otherwise granted in a separate license, the receiving Party's right to use, modify, or further transmit Digital Data is specifically limited to designing, constructing, using, maintaining, altering and adding to the Project consistent with the terms of this Exhibit, and nothing contained in this Exhibit conveys any other right to use the Digital Data.

§ 2.4 Where a provision in this Article 2 conflicts with a provision in the Agreement into which this Exhibit is incorporated, the provision in this Article 2 shall prevail.

ARTICLE 3 DIGITAL DATA PROTOCOLS

§ 3.1 Anticipated Types of Digital Data. The anticipated types of Digital Data to be used on the Project are as follows:
(Indicate below the information on the Project that shall be created and shared in a digital format. If the Parties indicate that Building Information Modeling will be utilized on the Project, the Parties shall also complete Article 4.)

Anticipated Digital Data	Applicability to the Project *(Indicate Applicable or Not Applicable)*	Location of Detailed Description *(Section 3.1.1 below or in an attachment to this exhibit and identified below)*
Project Agreements and Modifications		
Project communications		
Architect's preconstruction submittals		
Contract Documents		
Contractor's submittals		
Subcontractor's submittals		
Modifications		
Project payment documents		
Notices and claims		
Building Information Modeling		

§ 3.1.1 Insert a detailed description of the anticipated Digital Data identified in Section 3.1, if not further described in an attachment to this Exhibit.

§ 3.2 As soon as practical following execution of the Agreement, the Parties shall further describe the uses of Digital Data, and establish necessary protocols governing the transmission and Authorized Uses of Digital Data, in consultation with the other Project Participants that are expected to utilize Digital Data on the Project.

§ 3.2.1 Unless another Project Participant is identified below, the Architect shall prepare and distribute to the other Project Participants Digital Data protocols for review, revision and approval.
(If a Project Participant other than the Architect shall be responsible for preparing draft and final Digital Data protocols, identify that Project Participant.)

§ 3.2.2 The agreed upon Digital Data protocols shall be set forth in AIA Document G201–2013 and each Project Participant shall memorialize their agreement in writing to such Digital Data protocols.

§ 3.2.3 The Parties, together with the other Project Participants, shall review and, if necessary, revise the Digital Data protocols at appropriate intervals as required by the conditions of the Project.

§ **3.3** The Parties shall transmit, use, store and archive Digital Data in accordance with the Digital Data protocols set forth in the latest version of AIA Document G201–2013 agreed to by the Project Participants.

§ 3.4 Unauthorized Use
§ 3.4.1 Prior to Establishment of Digital Data Protocols
If a Party receives Digital Data prior to the agreement to, and documentation of, the Digital Data protocols in AIA Document G201–2013, that Party is not authorized to use or rely on the Digital Data. Any use of, or reliance on, such Digital Data is at that Party's sole risk and without liability to the other Party and its contractors, consultants, agents and employees.

§ 3.4.2 Following Establishment of Digital Data Protocols
Following agreement to, and documentation of, the Digital Data protocols in AIA Document G201–2013, if a Party uses Digital Data inconsistent with the Authorized Uses identified in the Digital Data protocols, that use shall be at the sole risk of the Party using the Digital Data.

§ 3.5 Digital Data Management
§ **3.5.1** Centralized electronic document management system use on the Project shall be:
(Check the appropriate box. If the Parties do not check one of the boxes below, the default selection shall be that the Parties will not utilize a centralized electronic document management system on the Project.)

☐   The Parties intend to use a centralized electronic document management system on the Project.

☐   The Parties do not intend to use a centralized electronic document management system on the Project.

§ **3.5.2** If the Project Participants intend to utilize a centralized electronic document management system on the Project, the Project Participants identified in Section 3.5.3 shall be responsible for managing and maintaining such system. The Project Participants responsible for managing and maintaining the centralized electronic document management system shall facilitate the establishment of protocols for transmission, use, storage and archiving of the centralized Digital Data and assist the Project Participants identified in Section 3.2.1 above in preparing Digital Data protocols. Upon agreement to, and documentation of, the Digital Data protocols in AIA Document G201–2013, the Project Participants identified in Section 3.5.3 shall manage and maintain the centralized electronic document management system consistent with the management protocols set forth in the latest version of G201–2013 approved by the Project Participants.

§ 3.5.3 Unless responsibility is assigned to another Project Participant, the Architect shall be responsible for managing and maintaining the centralized electronic document management system. If the responsibility for management and maintenance will be assigned to another Project Participant at an identified Project milestone, indicate below the Project Participant who shall assume that responsibility, and the Project milestone.
(Identify the Project Participant responsible for management and maintenance only if the Parties intend to utilize a centralized electronic document management system on the Project.)

Responsible Project Participant   **Project Milestone**

ARTICLE 4 BUILDING INFORMATION MODELING PROTOCOLS

§ 4.1 If the Parties indicate in Section 3.1 that Building Information Modeling will be used on the Project, specify below the extent to which the Parties intend to utilize Building Information Modeling and identify the provisions of this Article 4 governing such use:

☐ The Parties shall utilize Building Information Modeling on the Project for the sole purpose of fulfilling the obligations set forth in the Agreement without an expectation that the Model will be relied upon by the other Project Participants. Unless otherwise agreed in writing, any use of, transmission of, or reliance on the Model is at the receiving Party's sole risk. The remaining sections of this Article 4 shall have no force or effect.

☐ The Parties shall develop, share, use and rely upon the Model in accordance with Sections 4.2 through 4.10 of this Exhibit.

§ 4.2 Anticipated Building Information Modeling Scope. Indicate below the portions of the Project for which Modeling will be used and the anticipated Project Participant responsible for that Modeling.

Project Portion for Modeling   **Responsible Project Participant**

§ 4.3 Anticipated Model Authorized Uses. Indicate below the anticipated Authorized Uses of the Model for the Project, which Authorized Uses will be agreed upon by the Project Participants and further described for each LOD in G202–2013.

§ 4.4 Ancillary Modeling Activities. Indicate additional Modeling activities agreed upon by the Parties, but not to be included in AIA Document G202–2013, if any.
(Describe any Modeling activities, such as renderings, animations, performance simu-lations, or other similar use, including the anticipated amount and scope of any such Modeling activities.)

§ 4.5 Modeling protocols. As soon as practical following execution of the Agreement, the Parties shall, in consultation with the other Project Participants that are expected to utilize Building Information Modeling on the Project, fur-ther describe the Authorized Uses of the Model and establish necessary protocols governing the development of the Model utilizing AIA Document G202–2013.

§ 4.5.1 The Modeling protocols shall address the following:
.1 Identification of the Model Element Authors;
.2 Definition of the various LOD for the Model Elements and the asso-ciated Authorized Uses for each defined LOD;
.3 Identification of the required LOD of each Model Element at each identified Project milestone;
.4 Identification of the construction classification systems to be used on the Project;
.5 The process by which Project Participants will exchange and share the Model at intervals not reflected in Section 3.3, Model Element Table, of AIA Document G202–2013;
.6 The process by which the Project Participants will identify, coordi-nate and resolve changes to the Model;
.7 Details regarding any anticipated as-designed or as-constructed Authorized Uses for the Model, if required on the Project;
.8 Anticipated Authorized Uses for facilities management or other-wise, following completion of the Project; and
.9 Other topics to be addressed by the Modeling protocols: *(Identify additional topics to be addressed by the Modeling protocols.)*

§ 4.5.2 Unless responsibility is assigned to another Project Participant identified below, the Architect shall prepare and distribute Modeling protocols to the other Project Participants for review, revision and approval.
(If a Project Participant other than the Architect shall be responsible for preparing draft and final Modeling protocols, identify that Project Participant.)

§ 4.5.3 The agreed upon Modeling protocols shall be set forth in AIA Document G202–2013 and each Project Participant shall memorialize their agreement in writing to such Modeling protocols.

---

§ 4.5.4 The Parties, together with the other Project Participants, shall review, and if necessary, revise the Modeling protocols at appropriate intervals as required by the conditions of the Project.

§ 4.6 The Parties shall develop, use and rely on the Model in accordance with the Modeling protocols set forth in the latest version of AIA Document G202–2013, which document shall be included in or attached to the Model in a manner clearly accessible to the Project Participants.

§ 4.7 Unauthorized Use
§ 4.7.1 Prior to Establishment of Modeling protocols
If a Party receives any Model prior to the agreement to, and documentation of, the Modeling protocols in AIA Document G202–2013, that Party is not authorized to use, transmit, or rely on the Model. Any use, transmission or reliance is at that Party's sole risk and without liability to the other Party and its contractors, consultants, agents and employees.

§ 4.7.2 Following Establishment of Modeling protocols
Following agreement to, and documentation of, the Modeling protocols in AIA Document G202–2013, if a Party uses or relies on the Model inconsistent with the Authorized Uses identified in the Modeling protocols, such use or reliance shall be at the sole risk of the Party using or relying on the Model. A Party may rely on the Model Element only to the extent consistent with the minimum data required for the identified LOD, even if the content of a specific Model Element includes data that exceeds the minimum data required for the identified LOD.

§ 4.8 Model Management
§ 4.8.1 The requirements for managing the Model include the duties set forth in this Section 4.8. Unless assigned to another Project Participant, the Architect shall manage the Model from the inception of the Project. If the responsibility for Model management will be assigned to another Project Participant, or change at an identified Project milestone, indicate below the identity of the Project Participant who will assume that responsibility, and the Project milestone.

Responsible Project Participant   **Project Milestone**

§ 4.8.2 Model Management Protocol Establishment. The Project Participant responsible for managing the Model, in consultation with the other Project Participants that are expected to utilize Building Information Modeling on the Project, shall facilitate the establishment and revision of Model management protocols, including the following:

.1 Model origin point, coordinate system, precision, file formats and units
.2 Model file storage location(s)
.3 Processes for transferring and accessing Model files
.4 Naming conventions
.5 Processes for aggregating Model files from varying software platforms
.6 Model access rights
.7 Identification of design coordination and clash detection procedures.
.8 Model security requirements
.9 Other: (*Identify additional Model management protocols to be addressed.*)

§ **4.8.3 Ongoing Responsibilities**. The Project Participant responsible for managing the Model shall do so consistent with the Model management protocols, which shall also include the following ongoing responsibilities:

.1 Collect incoming Models:
    .1 Coordinate submission and exchange of Models
    .2 Create and maintain a log of Models received
    .3 Review Model files for consistency with Sections 4.8.2.1 through 4.8.2.5
    .4 Maintain a record copy of each Model file received.
.2 Aggregate Model files and make them available for Authorized Uses
.3 Maintain Model Archives and backups consistent with the requirements of Section 4.8.4 below
.4 Manage Model access rights
.5 Other: (*Identify additional responsibilities.*)

§ **4.8.4 Model Archives**. The individual or entity responsible for Model management as set forth in this Section 4.8 shall compile a Model Archive at the end of each Project milestone and shall preserve it without alteration as a record of Model completion as of that Project milestone.

§ **4.8.4.1** Additional Model Archive requirements, if any, are as follows:

§ **4.8.4.2** The procedures for storing and preserving the Model(s) upon final completion of the Project are as follows:

§ 4.9 Post Construction Model. The services associated with providing a Model for post construction use shall only be required if specifically designated in the table below as a Party's responsibility.

(Designate below any anticipated post construction Model and related requirements, the Project Participant responsible for creating or adapting the Model to achieve such uses, and the location of a detailed description of the anticipated scope of services to create or adapt the Model as necessary to achieve such uses.)

Post Construction Model	Applicability to Project *(Applicable or Not Applicable)*	Responsible Project Participant	Location of Detailed Description of Requirements and Services *(Section 4.10 below or in an attachment to this exhibit and identified below)*
§ 4.9.1 Remodeling			
§ 4.9.2 Wayfinding and Mapping			
§ 4.9.3 Asset/FF & E Management			
§ 4.9.4 Energy Management			
§ 4.9.5 Space Management			
§ 4.9.6 Maintenance Management			

§ 4.10 Insert a detailed description of the requirements for each Post Construction Model identified in Section 4.9 and the anticipated services necessary to create each Post Construction Model, if not further described in an attachment to this Exhibit.

ARTICLE 5 OTHER TERMS AND CONDITIONS

Other terms and conditions related to the transmission and use of Digital Data are as follows:

AIA® Document G201™ – 2013

Project Digital Data Protocol Form

PROJECT: *(Name and address)*

PROTOCOL VERSION NUMBER:
DATE:
PREPARED BY:
DISTRIBUTION TO: *(List each individual to whom this protocol is distributed. Include individuals listed in Section 1.2, or reference Section 1.2, along with any additional recipients.)*

TABLE OF ARTICLES

1   GENERAL PROVISIONS REGARDING USE OF DIGITAL DATA

2   DIGITAL DATA MANAGEMENT PROTOCOLS

3   TRANSMISSION AND USE OF DIGITAL DATA

ARTICLE 1 GENERAL PROVISIONS REGARDING USE OF DIGITAL DATA

§1.1 List each Project Participant that has incorporated AIA Document E203™– 2013, Building Information Modeling and Digital Data Exhibit, dated _____, into its agreement for the Project:

Project Participant	Discipline

§1.2 Project Participants. For each Project Participant listed in Section 1.1, identify and provide contact information for the individuals responsible for implementation of the Digital Data protocols.

Project Participant	Individual Responsible	Contact Information

§ **1.3** Terms in this document shall have the same meaning as those in AIA Document E203–2013.

ARTICLE 2 DIGITAL DATA MANAGEMENT PROTOCOLS

§ **2.1.1 Electronic Document Management System.** If, pursuant to Section 3.5.1 of the Project specific version of AIA Document E203–2013, the Project Participants indicated an intent to use a centralized electronic document management system on the Project, the requirements for the centralized electronic document management system are as follows:

(The requirements for the system shall address, among other things, access to and security of Digital Data.)

§ **2.1.2 System Startup Requirements.** Initial training and other startup requirements to be implemented with respect to the use or management of Digital Data, if any, are as follows:

(Describe in detail any initial training or other startup requirements.)

§ **2.1.3 Ongoing System Requirements.** Ongoing training or support programs to be implemented with respect to the use or management of Digital Data, if any, are as follows:

(Describe in detail any ongoing training or support programs to be implemented.)

§ **2.2 Digital Data Storage Requirements.** The procedures and requirements for storing Digital Data during the course of the Project, if any, are as follows:

(Describe in detail the procedures and requirements for storing Digital Data during the course of the Project.)

§ **2.3 Digital Data Archiving Requirements.** The procedures and requirements for archiving and preserving Digital Data during the course of the Project and following final completion of the Project, if any, are as follows:

(Describe in detail the procedures and requirements for archiving and preserving Digital Data during the course of the Project and following final completion.)

§ **2.4** Other Digital Data Management protocol requirements, if any, are as follows: *(Describe in detail any other requirements.)*

ARTICLE 3 TRANSMISSION AND USE OF DIGITAL DATA

§ **3.1 Digital Data Protocol Table**. The Project Participants shall comply with the data formats, transmission methods and Authorized Uses set forth in the Digital Data Protocol Table below when transmitting or using Digital Data on the Project. *(Complete the Digital Data Protocol Table by entering information in the spaces below. Adapt the table to the needs of the Project by adding, deleting or modifying the listed Digital Data as necessary. Use Section 3.2 Digital Data Protocol Table Definitions and Notes to define abbreviations placed, and to record notes indicated, in the Digital Data Protocol Table.)*

Digital Data		Digital Data Format	Transmission Method	Authorized Uses	Notes (*Enter #*)
§ 3.1.1	Project Agreements and Modifications				
§ 3.1.2	Project communications				
	General communications				
	Meeting notices				
	Agendas				
	Minutes				
	Requests for information				
	Architect's Supplemental Instructions				
§ 3.1.3	Architect's preconstruction submittals				
	Schematic Design Documents				
	Design Development Documents				
	Construction Documents				
§ 3.1.4	Contract Documents				
	Architect's Drawings				
	Architect's Specifications				
§ 3.1.5	Contractor's submittals				
	Product data				
	Submitted by Contractor				
	Returned by Architect				
	Shop drawings				

Digital Data	Digital Data Format	Transmission Method	Authorized Uses	Notes (*Enter #*)
Submitted by Contractor				
Returned by Architect				
§ 3.1.6 Subcontractor's submittals				
Product data				
Submitted by Subcontractor				
Returned by Contractor				
Shop drawings				
Submitted by Subcontractor				
Returned by Contractor				
§ 3.1.7 Modifications				
Requests for proposal				
Architect's order for a minor change in the Work				
Proposals				
Construction Change Directives				
Change Orders				
§ 3.1.8 Project payment documents				
§ 3.1.9 Notices and Claims				
§ 3.1.10 Closeout documents				
Record documents				
Operations and Maintenance Manual				

§ 3.2 Digital Data Protocol Table Definitions and Notes

Digital Data Format:
(Provide required data format, including software version, if applicable.)

Digital Data Format   Definition

Transmission Method:
(Below are suggested abbreviations and definitions. Delete, modify or supplement, as necessary.)

Abbreviation	Definition
CD	Delivered via Compact Disk
EM	Via e-mail
DMS	Centralized Electronic Document Management System

Authorized Uses of Digital Data:
(Below are suggested abbreviations and definitions. Delete, modify or supplement, as necessary.)

Abbreviation	Definition
I	Integrate (incorporate additional digital data without modifying data received)
M	Modify as required to fulfill obligations for the Project
R	Reproduce and distribute
S	Store and view only

Notes:
(List by number shown on table.)

AIA® Document G202™ – 2013

Project Building Information Modeling Protocol Form

PROJECT: *(Name and address)*

PROTOCOL VERSION NUMBER:
DATE:
PREPARED BY:
DISTRIBUTION TO: *(List each individual to whom this protocol is distributed. Include individuals listed in Section 1.1, or reference Section 1.1, along with any additional recipients.)*

This document is intended to be used in conjunction with a Project specific AIA Document E203™– 2013, Building Information Modeling and Digital Data Exhibit which the Parties will incorporate into their agreement for the Project, and a Project specific AIA Document G201™– 2013, Project Digital Data Protocol Form.

TABLE OF ARTICLES

1   GENERAL PROVISIONS

2   LEVEL OF DEVELOPMENT

3   MODEL ELEMENTS

ARTICLE 1 GENERAL PROVISIONS

§ 1.1 For each Project Participant that has incorporated the Project specific AIA Document E203™–2013, Building Information Modeling and Digital Data Protocol Exhibit, dated_____, into its agreement for the Project, identify and provide the contact information for individuals responsible for implementation of the Modeling protocols. If, for any Project Participant, more than one individual will be responsible for implementation of the Modeling protocols, list each individual separately and describe the unique Modeling Role assigned to each individual.

Modeling Role	Project Participant	Individual Responsible	Contact Information

§ 1.2 This document establishes the Modeling protocols for the Project. For purposes of these protocols, the Model is comprised of the following information and other data sets:
(Indicate disciplines, separate models, and other data that will be included within the Model and governed by the Modeling protocols.)

§ 1.3 Collaboration Protocols. The Project Participants' protocols for the collaborative utilization of the Model, if any, including communications protocols, a collaboration meeting schedule and colocation requirements, are as follows:

§ 1.4 Technical Requirements. The technical requirements relating to the utilization of Building Information Modeling, including specific software and hardware requirements are as follows:

§ 1.5 Training and Support. The parameters for any training or support program(s) that will be implemented with respect to any collaboration strategy or technical requirements are set forth below:

§ 1.6 Model Standard. The Model shall be developed in accordance with the following Model Standard, if any:

§ 1.7 Model Management Protocols and Processes
The following Model Management Protocols and Processes shall apply to the Project only if specifically designated in the table below as being applicable.
(Designate the Model Management Protocols and Processes applicable to the Project in the second column of the table below. In the third column, indicate whether the detailed description of the Model Management Protocol or Process is located in Section 1.8 or in an attached exhibit. If in an exhibit, identify the exhibit.)

Model Management Protocols and Processes	Applicability to Project *(Applicable or Not Applicable)*	Location of Detailed Description *(Section 1.8 below or in an attachment to this exhibit identified below)*
§ 1.7.1 Model origin point, coordinate system, precision, file formats and units		
§ 1.7.2 Model file storage location(s)		
§ 1.7.3 Processes for transferring and accessing Model files		

§ 1.7.4	Naming conventions		
§ 1.7.5	Processes for aggregating Model files from varying software platforms		
§ 1.7.6	Model access rights		
§ 1.7.7	Design coordination and clash detection procedures.		
§ 1.7.8	Model security requirements		

§ 1.8 Insert a description of each Model Management Protocol and Process identified in Section 1.7, if not further described in an exhibit attached to this document:

§ 1.9 Terms in this document shall have the same meaning as those in AIA Document E203-2013.

ARTICLE 2 LEVEL OF DEVELOPMENT

§ 2.1 The Level of Development (LOD) descriptions, included in Section 2.2 through Section 2.6 below, identify the specific minimum content requirements and associated Authorized Uses for each Model Element at five progressively detailed levels of completeness. The Parties shall utilize the five LOD descriptions in completing the Model Element Table at Section 3.3.

§ 2.2 LOD 100

§ 2.2.1 Model Element Content Requirements. The Model Element may be graphically represented in the Model with a symbol or other generic representation, but does not satisfy the requirements for LOD 200. Information related to the Model Element (i.e. cost per square foot, tonnage of HVAC, etc.) can be derived from other Model Elements.

§ 2.2.2 Authorized Uses

§ 2.2.2.1 Analysis. The Model Element may be analyzed based on volume, area and orientation by application of generalized performance criteria assigned to other Model Elements.

§ 2.2.2.2 Cost Estimating. The Model Element may be used to develop a cost estimate based on current area, volume or similar conceptual estimating techniques (e.g., square feet of floor area, condominium unit, hospital bed, etc.).

§ 2.2.2.3 Schedule. The Model Element may be used for Project phasing and determination of overall Project duration.

§ 2.2.2.4 Other Authorized Uses. Additional Authorized Uses of the Model Element developed to LOD 100, if any, are as follows:

§ 2.3 LOD 200

§ 2.3.1 Model Element Content Requirements. The Model Element is graphically represented within the Model as a generic system, object, or assembly with approximate quantities, size, shape, location, and orientation. Non-graphic information may also be attached to the Model Element.

§ 2.3.2 Authorized Uses

§ 2.3.2.1 Analysis. The Model Element may be analyzed for performance of selected systems by application of generalized performance criteria assigned to the representative Model Elements.

§ 2.3.2.2 Cost Estimating. The Model Element may be used to develop cost estimates based on the approximate data provided and quantitative estimating techniques (e.g., volume and quantity of elements or type of system selected).

§ 2.3.2.3 Schedule. The Model Element may be used to show ordered, time-scaled appearance of major elements and systems.

§ 2.3.2.4 Coordination. The Model Element may be used for general coordination with other Model Elements in terms of its size, location and clearance to other Model Elements.

§ 2.3.2.5 Other Authorized Uses. Additional Authorized Uses of the Model Element developed to LOD 200, if any, are as follows:

§ 2.4 LOD 300

§ 2.4.1 Model Element Content Requirements. The Model Element is graphically represented within the Model as a specific system, object or assembly in terms of quantity, size, shape, location, and orientation. Non-graphic information may also be attached to the Model Element.

§ 2.4.2 Authorized Uses

§ 2.4.2.1 Analysis. The Model Element may be analyzed for performance of selected systems by application of specific performance criteria assigned to the representative Model Element.

§ 2.4.2.2 Cost Estimating. The Model Element may be used to develop cost estimates suitable for procurement based on the specific data provided.

§ 2.4.2.3 Schedule. The Model Element may be used to show ordered, time-scaled appearance of detailed elements and systems.

§ 2.4.2.4 Coordination. The Model Element may be used for specific coordination with other Model Elements in terms of its size, location and clearance to other Model Elements including general operation issues.

§ 2.4.2.5 Other Authorized Uses. Additional Authorized Uses of the Model Element developed to LOD 300, if any, are as follows:

§ 2.5 LOD 400

§ 2.5.1 Model Element Content Requirements. The Model Element is graphically represented within the Model as a specific system, object or assembly in terms of size, shape, location, quantity, and orientation with detailing, fabrication, assembly, and installation information. Non-graphic information may also be attached to the Model Element.

§ 2.5.2 Authorized Uses

§ 2.5.2.1 Analysis. The Model Element may be analyzed for performance of systems by application of actual performance criteria assigned to the Model Element.

§ 2.5.2.2 Cost Estimating. Costs are based on the actual cost of the Model Element at buyout.

§ 2.5.2.3 Schedule. The Model may be used to show ordered, time-scaled appearance of detailed specific elements and systems including construction means and methods.

§ 2.5.2.4 Coordination. The Model Element may be used for coordination with other Model Elements in terms of its size, location and clearance to other Model Elements, including fabrication, installation and detailed operation issues.

§ 2.5.2.5 Other Authorized Uses. Additional Authorized Uses of the Model Element developed to LOD 400, if any, are as follows:

§ 2.6 LOD 500

§ 2.6.1 Model Element Content Requirements. The Model Element is a field verified representation in terms of size, shape, location, quantity, and orientation. Non-graphic information may also be attached to the Model Elements.

§ 2.6.2 Authorized Uses. Specific Authorized Uses of the Model Element developed to LOD 500, if any, are as follows:

ARTICLE 3 MODEL ELEMENTS
§ 3.1 Reliance on Model Elements
§ 3.1.1 At any particular Project milestone, a Project Participant may rely on the accuracy and completeness of a Model Element only to the extent consistent with the minimum data required for the Model Element's LOD for that Project milestone as identified below in the Model Element Table, even if the content of a specific Model Element includes data that exceeds the minimum data required for the identified LOD.

§ 3.1.2 Coordination and Model Refinement
Where conflicts are found in the Model, regardless of the phase of the Project or LOD, the Project Participant that identifies the conflict shall promptly notify the Model Element Authors and the Project Participant identified in AIA Document E203-2013 Section 4.8 as being responsible for Model management. Upon such notification, the Model Element Author(s) shall act promptly to evaluate, mitigate and resolve the conflict in accordance with the processes established in Section 1.7.7, if applicable.

§ 3.2 Table Instructions
§ 3.2.1 The Model Element Table in Section 3.3 indicates the LOD to which each Model Element shall be developed at each identified Project milestone and the Model Element Author.

§ 3.2.2 Abbreviations for each Model Element Author to be used in the Model Element Table are as follows:
(Provide abbreviations, such as "A—Architect," or "C—Contractor.")

Abbreviation	Model Element Author (MEA)		
§ 3.3 Model Element Table			
Identify (1) the LOD required for each Model Element at each Project milestone, (2) the Model Element Author (MEA), and (3) references to any applicable notes found in Section 3.4.			
Insert abbreviations for each MEA identified in the table below, such as "A—Architect," or "C—Contractor."			

Model Elements	Project Milestone 1			Project Milestone 2			Project Milestone 3			Project Milestone 4			Project Milestone 5			Project Milestone 6			Notes (See Sec 3.4)
	LOD	MEA	Notes	LOD	MEA	Notes	LOD	MEA	Notes	LOD	MEA	Notes	LOD	MEA	Notes	LOD	MEA	Notes	

§ 3.4 Model Element Table Notes

Notes:

(List by number shown on table)

Appendix B

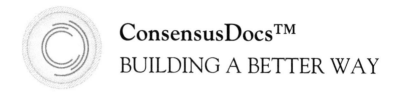

ConsensusDocs™
BUILDING A BETTER WAY

ConsensusDocs™ 301
Building Information Modeling (Bim) Addendum

General Instructions. These instructions are solely for the information and convenience of ConsensusDocs users, and are not a part of the document. Gray boxes indicate where you should click and type in your project information. The yellow shading is a Word default function that displays editable text and is not necessary for document completion. Shading can be turned off by going to the Review tab, select "Restrict Editing" button and uncheck "Highlight the regions I can edit." In Word 2003 you will find this option under the Tools tab, Options, Security tab, Protect Document button.

Embedded Instructions are provided to help you complete the document. To display or hide instructions select the "¶" button under the "Home" tab to show all formatting marks. Instruction boxes are color coded as follows:

Red Boxes: Instructions for fields that are typically required to complete contract.

Blue Boxes: Instructions for fields that may or may not be required for a complete contract.

Green Boxes: Provide general instructions or ConsensusDocs Coalition Guidebook comments, which can be found at www.Consensus Docs.org/guidebook.

Endorsement. This document was developed through a collaborative effort of organizations representing a wide cross-section of the design and construction industry. The organizations endorsing this document believe it represents a fair allocation of risk and responsibilities of all project participants.

Endorsing organizations recognize that this document must be reviewed and adapted to meet specific needs and applicable laws. This document has important legal and insurance consequences, and it is not intended as a substitute for competent professional services and advice. Consultation with an attorney and an insurance or surety adviser is strongly encouraged. Federal, State and Local laws may vary with respect to the applicability or enforceability of specific provisions in this document. CONSENSUSDOCS SPECIFICALLY DISCLAIMS ALL WARRANTIES, EXPRESS OR IMPLIED, INCLUDING ANY WARRANTY OF MERCHANTABILITY OR FITNESS FOR A PARTICULAR PURPOSE. PURCHASERS ASSUME ALL LIABILITY WITH RESPECT TO THE USE OF THIS DOCUMENT, AND CONSENSUSDOCS AND ANY OF THE ENDORSING ORGANIZATIONS SHALL NOT BE LIABLE FOR ANY DIRECT, INDIRECT OR CONSEQUENTIAL DAMAGES RESULTING FROM SUCH USE. For additional information, please contact ConsensusDocs, 2300 Wilson Blvd, Suite 400, Arlington, VA 22201, 866-925-DOCS (3627), support@consensusdocs.org or www.ConsensusDocs.org.

ConsensusDocs 301

Building Information Modeling (Bim) Addendum

General Principles

1.1. This Addendum does not effectuate or require a restructuring of contractual relationships or shifting of risks between or among the Project Participants other than as specifically required per the Addendum and its Attachments.

1.2. This Addendum is not intended to create privity of contract among any Project Participants beyond that which otherwise exists at law or by the terms of the Governing Contract.

1.3. Each Party to the Governing Contract shall append or incorporate, and shall cause each Project Participant with which it is in privity to append or incorporate, this identical Addendum in all contracts for which any other Project Participants are to perform obligations to be modeled. All such contracts shall contain flow-down provisions requiring that the provisions of this Addendum be passed downstream to subconsultants and subcontractors, as applicable.

1.4. Nothing in this Addendum shall relieve the Design Professional from its obligation, nor diminish the role of the Design Professional, as the person responsible for and in charge of the design of the Project.

1.5. Nothing in this Addendum shall diminish the extent to which, under applicable law, the Owner warrants to any Party the adequacy and/or sufficiency of the design.

1.6. Participation of the Contractor or its subcontractors and suppliers in Contributions to a Model shall not constitute the performance of design services.

1.7. Unless otherwise agreed in the BIM Execution Plan, a Design Model is not intended to provide the level of detail needed in order to extract precise material or object quantities.

1.8. In the event of a conflict between the contents of a Design Model and any other Model, the Design Model shall take precedence.

1.9. If any Project Participant becomes aware of a discrepancy between a Model and either another Model or another Contract Document, such Project Participant shall promptly notify the other Party or Parties to that Project Participant's Governing Contract and the Information Manager (IM).

1.10. Unless otherwise agreed in the BIM Execution Plan, the dimensional tolerances provided by the Contract Documents in the Governing Contract shall apply to dimensions in a Model.

1.11. In the event of an inconsistency between this Addendum and the Governing Contract, this Addendum shall take precedence.

2. Definitions

2.1. Affiliated Contract means any contract relating to the Project to which an identical Addendum is attached and in which that identical Addendum is incorporated, other than the Governing Contract.

2.2. Construction Model means a Model that (a) consists of those aspects of the Project that are to be modeled as specified in the BIM Execution Plan prepared pursuant to this Addendum; (b) utilizes data imported from a Design Model or, if none, from a designer's Construction Documents; and (c) contains the equivalent of shop drawings and other information useful for construction.

2.3. Contract Documents, as defined in the Governing Contract, is modified to include all Design Models, unless otherwise specified in the BIM Execution Plan.

2.4. Contribution means the expression, design, data or information that a Project Participant (a) creates or prepares, and (b) incorporates, distributes, transmits, communicates or otherwise shares with other Project Participant(s) for use in or in connection with a Model for the Project.

2.5. Contributor means a Project Participant who makes a Contribution.

2.6. Design Model means a Model of those aspects of the Project that (a) are to be modeled as specified in the BIM Execution Plan prepared pursuant to this Addendum and (b) have reached the stage of completion that would customarily be expressed by a Design Professional in two-dimensional Construction Documents. This shall not include Models such as analytical evaluations, preliminary designs, studies, or renderings. A Model prepared by a Design Professional that has not reached the stage of completion specified in this definition is referred to as a Model.

2.7. Drawings means (a) those two-dimensional plans, sketches or other drawings that are Contract Documents under the Governing Contract and are created separately from, and are not derived from, a Model and (b) those two-dimensional projections derived from a Model supplemented with independent graphics and annotations specified by the Parties to be Contract Documents.

2.8. Federated Model means a Model consisting of linked but distinct component Models, drawings derived from the Models, texts, and other data sources that do not lose their identity or integrity by being so linked, so that a change to one component Model in a Federated Model does not create a change in another component Model in that Federated Model.

2.9. Full Design Model means a Model consisting of coordinated structural, architectural, MEP and other Design Models designated in the BIM Execution Plan to be produced by the design team.

2.10. Governing Contract means the agreement to which this Addendum is attached and in which it is incorporated, but excludes an Affiliated Contract.

ConsensusDocs™ 301 - Building Information Modeling (BIM) Addendum - © 2008, Revised 2011. THIS DOCUMENT MAY HAVE BEEN MODIFIED FROM THE STANDARD LANGUAGE, and a report of modifications can be generated through the ConsensusDocs platform. Purchase of the document permits the user to print one contract for each party to the contract within one project only. You may only make copies of finalized documents for distribution to parties in direct connection with this contract. Any other uses are strictly prohibited.

2.11. Information Management means measures that protect and defend information and information systems with respect to their availability, integrity, authentication, confidentiality, and nonrepudiation. These measures include providing for restoration of information systems by incorporating protection, detection, and reaction capabilities.

2.12. Information Manager or IM means one or more individuals responsible for the BIM's Information Management program.

2.13. MEP means mechanical, electrical and plumbing.

2.14. Model means a three-dimensional representation in electronic format of building elements representing solid objects with true-to-scale spatial relationships and dimensions. A Model may include additional information or data.

2.15. Project Model means a Model consisting of the federation of a Full Design Model and one or more Construction Models designated in the BIM Execution Plan to be produced by Project Participants.

2.16. Project Participant shall be, and Project Participants shall include, each Party to the Governing Contract and each Party to an Affiliated Contract.

3. Information Management

3.1. The Owner or its designated representative shall appoint one or more IM(s) for the Project. Unless otherwise agreed upon in the BIM Execution Plan, all compensation and related costs for the Information Manager(s) shall be paid by the Owner. The Project Participant whose box is checked as indicated below shall serve as IM until replaced. The Owner may replace the IM at its own discretion.

[   ] Design Professional
[   ] Contractor/Construction Manager
[   ] Other [   ] (specify).

3.2. The role and responsibility of the IM with respect to a Federated Model for the Project, including the Project Model, shall be to perform or procure from a third party acceptable to the Owner the following functions (exclude any functions that do not apply):

3.2.1. Create, delete, modify and maintain user accounts;

3.2.2. Assign, delete and modify access rights to users;

3.2.3. Apply access controls to users so that only authorized users of the Model can access only the data they are authorized to access;

3.2.4. If appropriate, establish and maintain encryption-at-rest measures and encryption-during-transmissions measures;

3.2.5. Record, at a minimum, the following information about each data entry by Model users in the Federated Model (including downloading of Models to the Federated Model):

   a) User name;
   b) User role;
   c) Contact information;
   d) Date/time entered;
   e) Any additional information required to be recorded for each data entry as set forth in the BIM Execution Plan;

3.2.6. Backup and restore data;

3.2.7. Routinely run information system scans to maintain Model security;

3.2.8. Maintain and monitor information system logs so that only authorized users are accessing the Model and to ensure that there are no functional problems associated with the Model;

3.2.9. Install patches to close documented vulnerabilities in the Model;

3.2.10. Document and report any incident relating to the Model (including but not limited to an incident originating outside the Model that results in the Model being the victim of an attack) and take action to protect the Model;

3.2.11. Transfer unconditionally to a successor IM, at such times as directed by the Owner, all tangible and intangible property and information that came into its possession, custody or control in its capacity as IM;

3.2.12. Provide authorized users with access instructions and system requirements;

3.2.13. Respond to requests by authorized users for assistance in maintaining access; and

3.2.14. Perform any and all other responsibilities or functions as required of it in the BIM Execution Plan.

4. BIM Execution Plan

4.1.   As soon as is practicable, but in no event later than thirty (30) days after the latter of the execution of the Contract between the Owner and the Design Professional or execution of the Contract between the Owner and the Contractor or Construction Manager, all Project Participants shall meet, confer and use their best efforts to agree upon the terms of or modifications to a BIM Execution Plan. When agreed upon, the BIM Execution Plan and any modifications shall become an amendment to this Addendum. As soon as is practicable, but in no event later than thirty (30) days after the execution of a Contract with any other Project Participants, all Project Participants shall meet, confer and use their best efforts to agree upon any necessary modifications to a BIM Execution Plan.

4.2.   Unless otherwise agreed, the IM shall schedule and chair all such meetings.

4.3.   The BIM Execution Plan shall address the following elements, but may include additional elements:

4.3.1. Contact information for each Project Participant;

4.3.2. Identification of what Models are to be created, the purpose(s) each Model is intended to serve, and which Project Participant(s) is(are) responsible for creating each Model;

4.3.3. A definition of what Design Model or Models, if any, shall not constitute Contract Documents;

4.3.4. The spatial portions or areas of the Project to be modeled in each Model and the spatial portions or areas of the Project not to be modeled;

4.3.5. The expected content of each Model and the required level of detail at various Project milestones, which content includes:

    a. geometric and spatial data;
    b. object property data;
    c. object constitution data;
    d. provision for object parameters as place holders for cost and schedule data; or
    e. authoritative source information;

4.3.6. A schedule for initial delivery of each Model to the IM;

4.3.7. A schedule for updating of each Model and preservation of versions of each Model and its constituent Models;

4.3.8. A definition of what Model or Models shall constitute part of the record documents for the Project;

4.3.9. Procedures and protocols for submission, for approval of Models including electronic stamping, for designating a Model as a Design Model, and for notification of action on a request for approval;

4.3.10. Procedures and protocols for designating two-dimensional projections derived from a Model as Contract Documents;

4.3.11. Contributor's Dimensional Accuracy Representation Selection of one, but only one, of the following representations applicable to the dimensional accuracy of any Contribution of or to a Model. Any such representation is:

    1. limited to the other parties to the Governing Contract,
    2. in accordance with the standard of care applicable to the Contributor for such Contribution, and
    3. effective at the time the Model has been developed to the same stage of completion as two-dimensional Construction Documents.

[   ] Each Contributor represents that the dimensions in its Contribution to a Model are accurate and take precedence over the dimensions called out in the Drawings or inferred from the Drawings. Details and components that are not represented in a Contribution to a Model must be retrieved from the Drawings;

[   ] Each Contributor represents that the dimensions in its Contribution to a Model are accurate to the extent that the BIM Execution Plan specifies dimensions to be accurate, and all other dimensions must be retrieved from the Drawings;

[   ] Contributors make no representation with respect to the dimensional accuracy of the Contributor's Contribution to a Model. A Model can be used for reference only and all dimensions must be retrieved from the Drawings; or

[   ] Other: [   ];

4.3.12. Establishment of a common coordinate system;

4.3.13. Establishment of conventions as to units;

4.3.14. Conventions for defining critical dimensions and critical Model content;

4.3.15. File format to be used;

4.3.16. File-naming and object-naming conventions to be used;

4.3.17. File structure to be used;

4.3.18. Software to be utilized;

4.3.19. Measures needed to achieve interoperability of applications;

4.3.20. Two-dimensional reference Drawings;

4.3.21. Utilization of BIM for the RFI process, response protocol and timing, incorporation of responses into any Model;

4.3.22. Utilization of BIM for the Change Order process, response protocol and timing, incorporation of responses into any Model;

4.3.23. A schedule for BIM development, coordination and clash detection meetings among the Project Participants;

4.3.24. Engagement of the IM in these processes;

4.3.25. Utilization of a Project BIM website;

4.3.26. Procedures and protocols for confirmation of field changes through an as-built Project Model;

4.3.27. Specification of Project close-out and final deliverables;

4.3.28. The extent, if any, to which Project Participants or specified staff for each will be co-located; and

4.3.29. Any changes or additions to the Governing Contract or an Affiliated Contract relating to BIM-related compensation and costs.

5. Risk Allocation

5.1. Each Party shall be responsible for any Contribution that it makes to a Model or that arises from that Party's access to that Model. Such responsibility includes any Contribution or access to a Model by a Project Participant in privity with that Party and of a lower tier than that Party. Nothing in this section shall expand the scope of any representation stated in the BIM Execution Plan pursuant to Section 4.3.11.

5.2. With respect to the issue of a waiver of consequential damages:

  a. The Governing Contract shall govern the issue of any waiver of consequential damages arising from a Contribution; and

  b. Each Party waives claims against the other Parties to the Governing Contract for consequential damages arising out of or relating to the use of or access to a Model, including but not limited to damages for loss of use of the Project, rental expenses, loss of income or profit, costs of financing, loss of business, principal office overhead and expenses, loss of reputation or insolvency.

5.3. To the extent that any or all Design Models are included as Contract Documents, Project Participants may rely upon the accuracy of information in those Design Models; provided, however, that regardless of whether any Design Models are included as Contract Documents, the selection in Section 4.3.11 shall control a Project Participant's right to rely on the dimensional accuracy of a Contribution or Model.

5.4. The standard of care applicable to each Party regarding that Party's Contributions to or use of a Model shall be in accordance with that Party's Governing Contract or common law, as applicable.

5.5. Each Party shall use its best efforts to minimize the risk of claims and liability arising from the use of or access to its Model or the Project Model. Such efforts shall include promptly reporting to the relevant Project Participants any errors, inconsistencies, or omissions it discovers in its Model or the Project Model; however, nothing in this section shall relieve any Party of liability it would otherwise bear under Section 5.1.

5.6. No Party involved in creating a Model shall be responsible for costs, expenses, liabilities, or damages which may result from use of its Model beyond the uses set forth in this Addendum or fully executed amendments hereto.

5.7. Unless agreed otherwise in the BIM Execution Plan, each Party shall (a) procure and maintain valuable papers and records insurance coverage, with limits no less than $[   ], covering all of the Party's Contributions or intended Contributions; (b) include this requirement in its contracts with any other Project Participant; and (c) provide the other with a certificate of insurance demonstrating compliance with this requirement by that Party and such other Project Participants.

5.8. A defect in the software used in the creation, modification, federation or other use of a Model, including the Project Model, shall entitle a Party to a time extension or other excuse from performance, but only to the extent that the Party could not have avoided any delay or loss by the exercise of reasonable care. In addition, a Party has the duty to mitigate any such delay or loss.

6. Intellectual Property Rights In Models

6.1. Each Party warrants to the other Parties to the Governing Contract that either (a) that Party is the owner of all copyrights in all of that Party's Contributions, or (b) that Party is licensed or otherwise authorized by the holders of copyrights of expression contained in the Contribution to make such Contribution under the terms of this Addendum. Subject to waiver of subrogation clauses, if any, contained in the Governing Contract, each Party agrees to indemnify and hold such other Parties harmless for claims of third parties arising out of, or relating to, claims or demands relating to infringement or alleged infringement of expression contained in that Party's Contribution. Nothing in this Addendum is intended to limit, transfer, or otherwise affect any of the intellectual property rights that a Party may have with respect to any Contribution, except for the licenses or permissions expressly granted by this Addendum or the Governing Contract.

6.2. Subject to the provisions of Section 6.1, each Party grants to the other Party or Parties to the Governing Contract (a) a limited, non-exclusive license to reproduce, distribute, display, or otherwise use that Party's Contributions for purposes of this Project only; (b) a limited, non-exclusive sublicense to reproduce, distribute, display, or otherwise use, for purposes of this Project only, the Contributions of those other Project Participants who have granted that Party an identical license or sublicense; (c) the right to grant an identical sublicense to any other Project Participants with which the licensee has an Affiliated Contract in which this Addendum is incorporated by reference; and (d) a limited, non-exclusive license to reproduce, distribute, display, or otherwise use any Model containing such Contributions, or any other Model with which the Model containing such Contributions is federated or otherwise related, in each case for the sole purpose of carrying out the Project Participants' respective duties and obligations relating to this Project. This limited license shall include any archival purposes permitted hereunder or in the Governing Contract, but does not allow the licensee to reproduce, distribute, display, or otherwise reuse all or part of any other Party's Contributions except as permitted herein or in the Governing Contract. This limited, non-exclusive license is in addition to any other licenses or usage rights that also may be granted under the Governing Contract.

6.3. If a Party to the Governing Contract is the holder of copyrights in another Project Participant's Contribution or is the grantee of an exclusive license with respect to such Contribution, then such holder or exclusive licensee hereby grants to such other Party or Parties the right to grant to other Project Participants with which the other Party has or Parties have Affiliated Contracts in which this Addendum is incorporated, a limited license in the form set forth in Section 6.2.

6.4. The Project Owner's entitlement to use the Full Design Model after completion of the Project shall be governed by the Contract between the Owner and the Design Professional.

6.5. Unless otherwise limited herein or by express license-limiting terms in the Governing Contract, the non-exclusive license granted in this BIM Addendum shall remain in effect as permitted by law. In addition, after final completion of the Project, the non-exclusive license shall be limited to keeping an archival copy of Project-related Contributions.

6.6. In the absence of express language to the contrary in the Governing Contract or in this Addendum, nothing in this Addendum, and no act by a Project Participant in furtherance of this Addendum, shall be deemed or construed to deprive or dispossess a Contributor of copyrights or license rights held by that Contributor in its respective underlying Contribution to any Model. Other Parties, Project Participants, persons, or entities that provide Contributions to a Model shall not be deemed to be co-authors in the Contributions of other Project Participants. Except where otherwise stated, no Contributor shall possess rights in a Model containing a Contribution greater than those granted by the non-exclusive license provided in this Addendum, as that license may be further limited in this Part 6. Nothing in this Addendum shall grant a right to a Party to use all or part of another Party's Contribution for any purpose other than performance of the Project Participant on this Project and as is otherwise expressly stated in the Governing Contract or in this Addendum.

6.7. Terms of the Governing Contract pertaining to non-payment by the Project Owner notwithstanding, the Project Owner's non-exclusive license granted herein to reproduce, distribute, display, or otherwise reuse the Contributions and Models shall not be limited to construction or maintenance of this Project. However, if the Project Owner fails materially in its Project-related payment obligations to a Contributor, and that material failure is so adjudged against the Project Owner by the decision of a court of law or arbitration (an Adjudication), then any Project-related licenses to the Project Owner from that Contributor shall be terminated as of the time of such Adjudication.

Notwithstanding the foregoing, the Parties hereto (and all Contributors by virtue of Affiliated Contracts) waive any rights to claim contributory, direct, or vicarious copyright infringement, to assert claims of misappropriation or like claims, to revoke licenses granted herein, or to pursue equitable remedies under the Copyright Act or under applicable law against other Parties and Contributors who are not found liable in the Adjudication (for failure to pay or otherwise). This applies to their respective obligations, if any, to the Contributor unpaid by the Project Owner, and the non-liable Parties' licenses granted herein shall survive the termination of the Owner's license due to the Adjudication.

Index

acceleration: BIM, as means to better manage *see also* BIM: general overview 22–36; constructive, as type 8; damages sought 8; damages, in comparison with delay claim damages 9; directed, as type 8

AIA *The Architect's Handbook of Professional Practice*: treatment of Building Information Modeling (BIM) and Computer Aided Design (CAD), 10th through 14th editions 65

American Institute of Architects (AIA) *CAD Layer Guidelines* 63

AIA document A201: contractor licensing requirements 46; drawings, defined 114; time extension for excusable delay 10; revision cycles, historically 65

AIA document B101: BIM, listed as potential additional service 115; construction documents, defined 85; instruments of service, licensing of 68; mutual waiver of consequential damages 12; standard of care, defined 45

AIA document B141: standard of care, implied 45; reliance, as model definition in *Taylor* case 110

AIA document B151: standard of care, implied 45

AIA document C106 64

AIA document C131: workmanlike skill and care of construction managers 46

AIA document E202 *see* BIM form (boiler-plate) contracts

AIA document E203 *see* BIM form (boiler-plate) contracts

AIA document G201 *see* BIM form (boiler-plate) contracts

AIA document G202 *see* BIM form (boiler-plate) contracts

architectural drawings, historical perspective 120

as-built construction BIM 31, 49, 53–54, 69; facilities management data, as source of 77; AIA document G202 81; ConsensusDocs document 301 88

Associated General Contractors (AGC): *Contractors' Guide to BIM* [*see also* BIM guidelines and standards]; BIM certificate program 144; BIMForum *LOD Specification* 69

automatic updates between BIM and two-dimensional drawings *see* bi-directional associativity

barcode 34

bi-directional associativity: general concept 24–31; possible argument for BIM as contract document 115

BIM360 Glue 30, 32, 35

BIM360 Field 34

Building Information Modeling (BIM): general overview 22–36; construction BIM 31–36; construction BIM software 31–34; construction BIM processes 34–36; definitions of native and compiled BIMs 25; design BIM processes 27–31; design BIM software 26–31; development, use, and reliance 109–114; legal status 114–117; relationship between BIM and 2D contract drawings 24

BIM execution plan (BEP) *see* BIM form (boiler-plate) contracts, esp. ConsensusDocs 301 (2008) 83–92

BIM form (boiler-plate) contracts: AIA E-202 (2008) 64–71; AIA E203 (2013) 71–78; AIA G201 (2013) 78–79; AIA G202 (2013) 79–83; ConsensusDocs 301 (2008) 83–92

BIM guidelines and standards: AGC Contractors Guide to BIM 50; Indiana University 52–54; State of Wisconsin 54–56; US Dept. of Veterans Affairs 56–58; State of Ohio 58–59; New York City 59–62; National Building Information Modeling Standard 62–63

Center for Integrated Facility Engineering (CIFE) 48–9

change order: BIM planning questions 142; BIM requirements in ConsensusDocs document 301 90; claims documents evaluation process 135–38; errors and omissions, in context of 44; generally in claims processes 6–10; reductions in quantity as a result of BIM, per Ohio State Architect Office research 58; contract clause example in ConsensusDocs document 200 73;

claims: *see also* acceleration; *see also* contractor claims against owners; *see also* delay; *see also* disruption; *see also* owner claims against contractors; *see also* scope changes

clash detection/reporting: AIA document E202, requirements in 69; AIA document G202, requirements in 82; CIFE, early research 49; generally 23–33; Indiana University BIM Guidelines & Standards for Architects, Engineers and Contractors, requirements in 53; New York City Department of Design and Construction (DDC) BIM guidelines, requirements in 62; standard of care and workmanlike performance, suggested minimums 92, 134; trade coordination, typical hierarchy 34; Veterans Affairs BIM Guide 57

code checking automation 26, 60, 117

commissioning: *see also* COBie; *see also* facilities management; Veterans Affairs BIM Guide 57

Computer Aided Design (CAD) as distinct from BIM 22, 24–31

conformed design BIM 54, 59, 77

construction contracts, modern complexity of 4

Construction Operations Building Information Exchange (COBie): generally 54; Indiana University BIM Guidelines & Standards for Architects, Engineers and Contractors 54; Veterans Affairs BIM Guide 57

Construction Specification Institute (CSI): MasterFormat (1995 and 2004) classification system, for keynote annotation in Revit 33; Uniformat classification system, [for model objects in Revit 33]; [as schema for AIA document E202 Model Element Table 66]; [removed in AIA document G202 83, 113]; *Uniform Drawing System* 63

ConsensusDocs document 200: change orders, 72; excusable delay 9; workmanlike performance as distinct from licensed professional services 46

ConsensusDocs document 240: standard of care, described 45

ConsensusDocs document 301 *see* BIM form (boiler-plate) contracts

consequential damages: waiver of, in AIA document B101 12; waiver of, in ConsensusDocs 301 90–1, 123

contractor claims against owners 4–5

contract documents: 2D extractions from BIM 24, 28, 86, 117–20; BIM, as 114–17; coordination, as improved by BIM 30; defective and deficient 6–7, 24; precedence over BIM, in cases of discrepancy 31, 59, 91; *see also Spearin* warranty

copyright and intellectual property 73, 91, 124–126

damages: BIM process, as type in delay and disruption 138; discrete cost; 131; equipment costs 12; financing 10, 12, **18**; labor costs 8–12, 131; material costs 10, 12, 131; modified total cost 131; overhead costs 10, 12, 90; quantum meruit 131; total cost approach 131

data loss and archiving 123–4
data management: contract
  requirements in AIA document
  E203 74; G201 78–79, 136; generally
  75–77;
delay: *compensable* 10; *excusable* 8, 10;
  damages sought 9–10; "no damages for
  delay" clauses 11; *non-compensable* 10,
  non-excusable 10
design–bid–build (DBB) 1,56, 57, 61
design–build (DB) 1,56, 57, 58
disruption 11–12, **13**,17, **18**, 36,
  137, 138
document management *see* data
  management
drawing eXchange format (DXF) 120
duty to inquire 17, 32, 36, 67, 82

EastCoastCAD 32
economic loss rule 14–17
Eichleay: 10
errors and omissions 7, 41, 144
estimating: authorized use of BIM in AIA
  documents E202 and G203 111 14, 31,
  33, 69, 70, 76, 80; software errors, in
  Mortenson case 91; 111
expert opinion and testimony 47

facilities management: compensation
  for BIM development towards
  77; cost analysis, NIST report on
  interoperability 49; CIFE research,
  examples 49–50; generally 22–24; BIM,
  license for use post construction 68,
  76; BIM guidelines [Indiana University
  BIM Guidelines & Standards for
  Architects, Engineers and Contractors
  54] [Veterans Affairs BIM Guide 57–8];
  owner's data ownership in BIM, State
  of Ohio Building Information Modeling
  Protocol 126;
FIATECH 117
five-dimensional (5D) 33
four-dimensional (4D) 33, 134

implied warranty: 4, 7, 14, 17, *see also*
  Spearin warranty
Industry Foundation Class (IFC) 56,
  121–122

insurance: builders risk 44; commercial
  general liability 44, 51; professional
  liability 44, 51, 137
Integrated Project Delivery (IPD) 3,
  53, 61
interoperability 120–122

laser scanning 61
Level of Development (LOD) *see also* AIA
  document E202 *see also* AIA document
  G202 *see also* AGC [BIMForum *LOD
  Specification* 69]
liquidated damages 12

measured mile calculation 132

National Building Information Modeling
  Standard (NBIMS) 22, 24, 62–63,
  104
National CAD Standard 63
National Council of Architectural
  Registration Boards (NCARB): intern
  development program (IPD), possible
  changes as result of technology 92,
  108–9, 134, 144; responsible control,
  defined in *Rules of Conduct* 106
National Institute of Building Sciences
  (NIBS) *Plotting Guidelines* 63
National Institute of Standards and
  Technology (NIST) 49, 62
National Society of Professional
  Engineers 106
Navisworks 30–36, 53, 122
negligence 8, 12, 14, **18**, 47
negligent misrepresentation 14–17, **18**,
  138
Mechanical Contractors Association of
  America (MCAA) *Change Orders,
  Productivity, Overtime: A Primer for the
  Construction Industry* 12

pre-fabrication 11, 32,
privity of contract 13–17, 84, 90,
  113, 123
professional licensing 17, 92, 108, *see also*
  standard of care
Public Private Partnership (P3) 3

Quickpen 32

requests for information (RFI) 25, 50, 90,
135–36
responsible control 56, 83, 86, 104–109,
113–17
Restatement (Second) of Torts 14–18, 138
Revit 26–35; 52, 119, 135

scope changes 5–7: directed changes 5–6;
constructive change 6
software version control 122–123
Solibri Model Checker 26; *see also* code
checking automation
Spearin warranty 7, 17, 56, 84, 86, **105**,
SprinkCAD 32
standard of care: general concepts 43–47;
evolution of BIM/VDC documented
in: general research 48–51; BIM
guidelines and standards 52–63; AIA
and ConsensusDocs form (boiler-plate)
contracts 63–93; emerging minimum
expectations 92, 134
subcontractor: BIM workflows, typical
31–36, 51; BIM not requiring
performance of design services 86;

recipient of computerized schedule in
RW Vaught 104; responsibility for BIM
contributions 90, 107; BIM as potential
demonstrative in labor inefficiency
claim 136
surety 51, 144

Tekla 32
torts: 12–17;
two-dimensional (2D) CAD – three
dimensional (3D) conversions 117–120

Vitruvius *see* architectural drawings
Virtual Design and Construction (VDC)
see Building Information Modeling
(BIM)

workmanlike performance: general
concepts 43–47; evolution of BIM/
VDC documented in: general research
48–51; BIM guidelines and standards
52–63; AIA and ConsensusDocs form
(boiler-plate) contracts 63–93; emerging
minimum expectations 92, 134

For Product Safety Concerns and Information please contact our EU
representative GPSR@taylorandfrancis.com
Taylor & Francis Verlag GmbH, Kaufingerstraße 24, 80331 München, Germany

www.ingramcontent.com/pod-product-compliance
Ingram Content Group UK Ltd.
Pitfield, Milton Keynes, MK11 3LW, UK
UKHW021120180425
457613UK00005B/162